NASHUA COMMUNITY COLLEGE

The
WEIRD & WONDROUS
WORLD of PATENTS

Robert O. Richardson

D1405089

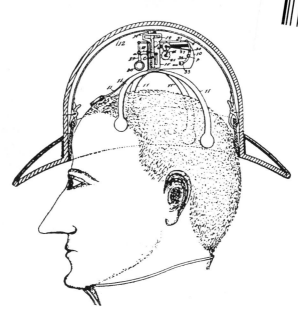

Sterling Publishing Co., Inc.   New York

To my wife, Viola,
in appreciation
for her help and tolerance
in putting this book together.

Edited by Timothy Nolan
Designed by Lisa Krivacka and Timothy Nolan

Library of Congress Cataloging-in-Publication Data

Richardson, Robert O.
    The weird & wondrous world of patents / by Robert O. Richardson.
        p.   cm.
    Includes index.
    1. Patents—Miscellanea.   2. Inventions—Miscellanea.   I. Title.
  II. Title: Weird and wondrous world of patents.
  T215.R53   1990
    608—dc20   90-39882
                                                                CIP

10   9   8   7   6   5   4   3   2   1

© 1990 by Robert O. Richardson
Published by Sterling Publishing Company, Inc.
387 Park Avenue South, New York, N.Y. 10016
Distributed in Canada by Sterling Publishing
⅝ Canadian Manda Group, P.O. Box 920, Station U
Toronto, Ontario, Canada M8Z 5P9
Distributed in Great Britain and Europe by Cassell PLC
Villiers House, 41/47 Strand, London WC2N 5JE, England
Distributed in Australia by Capricorn Ltd.
P.O. Box 665, Lane Cove, NSW 2066
Manufactured in the United States of America
All rights reserved

Sterling ISBN 0-8069-7250-5

CONTENTS

N. H. TECHNICAL COLLEGE AT NASHUA
T215.R53 1990
Richardson, The weird & wondrous world

3 0426 0000 7267 9

ABOUT THE AUTHOR

Robert O. Richardson, a former Patent Office Examiner, holds a Doctor of Laws degree from George Washington University, is listed in *Who's Who in American Law*, and has been a registered patent attorney for over 35 years. He has represented the United States Navy, the United States Army, and a number of corporations, as well as a host of individual investors. All told, he counts over 500 successful patent prosecutions to his credit.

INTRODUCTION

History of Patents

Science students are quick to point out that during the millions of years that humans have been on earth, only in the last thousand years has technology skyrocketed. Starting with the Industrial Revolution, governments realized the importance of research and development of machines, materials, and methods of manufacture to improve the lives of their subjects. At one time, kings, monarchs and other rulers would award monopolies to friends and other "worthy" applicants to exclusive use of government properties, such as exclusive hunting rights in government forests, for example. Gradually pressure was applied to cause monopoly recipients to earn these rights. Eventually, if someone made an improvement or invention on a device, he earned the right to exclusive use of it for a period of time before the public could have free use of it. Thus, the patent system was born.

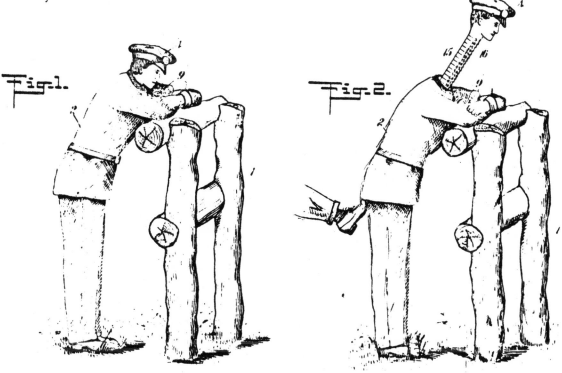

How many times have you wanted to do this?

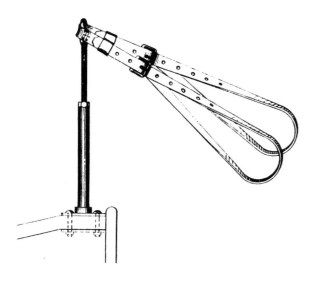

Article 1, Section 8, of the American Constitution gives Congress the right to award writers and inventors for their writings and discoveries. Thus, patent and copyright laws find their basis in the Constitution. Our forefathers realized the importance of incentives in developing inventions, so patents give inventors the right to exclude others for a limited time from making, using or selling that which is invented, as set forth in the patent. This is one reason why the United States, in its very short existence, leads older countries that do not offer such incentives in patentable inventions.

In 1790 President Washington urged Congress to effectively encourage the exertion of skill and genius. The first Patent Commission included Secretary of State Thomas Jefferson, Secretary of War Henry Knox and Attorney General Edmund Randolph, and granted the first patent in 1790 to Samuel Hopkins for an improvement in the making of potash, an ingredient in lye soap. The Patent Bill of 1793 abolished the examination system in favor of a simple registration system. However, from 1793 to 1836, this system produced worthless and void patents, many in conflict with each other. (Dr. William Thornton, a close friend of James Madison, served as Patent Chief from 1802 to 1828. His getting several patents himself led to the prohibition of Patent Office employees from getting patents.) The Patent Act of July 4, 1836 reestablished the examination system as necessary for

Necessity is the mother of invention, or in the case of this patent, the schoolmarm.

determining the utility and novelty of an invention, and required examiners to search for pre-existing patents or inventions. It also started the present numbering system. After two fires in 1836 and 1877 the submission of models was no longer required, except in unusual cases when the examiner needed one for a better understanding of the invention.

The Design Patent Act passed Aug. 29, 1842 established the *design patent* system in which artistic configurations pleasing to the eye were protected if sufficiently novel. Design Patent No. 1 was issued in 1842 to George Bruce for a new and original design for printing type. Design patents have a term of 14 years.

Since 1930 *plant patents* have been granted to persons who have invented and asexually reproduced any distinct and new variety of plant. Luther Burbank and Thomas Edison urged that farmers and plant breeders should have the same status as other inventors.

Prior to 1861 patents were granted for 14 years, extendable for another 7 years. In 1861 the term was changed to 17 years. Today, to keep the patent in force for the full term, maintenance fees must be paid 3½, 7½ and 10½ years after the issuance date.

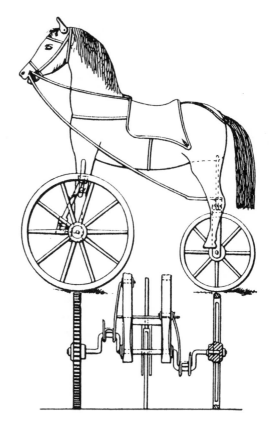

Self-propulsion was an obsession for inventors in the 19th century, but this patent for a "velocipede" was truly a horse of another color.

Patent Law

Basically, a patent is simply a contract between a government and the inventor, giving the inventor the right to exclude others from using the invention (as claimed) for a limited time. In return, the inventor must disclose to the public enough information about the invention so that when the patent term expires, the public can use it. This is why the invention is shown in drawings and described in sufficient detail that it can be understood and reproduced with a minimum of experimentation. The patent doesn't give the inventor the right to use his own invention, merely the right to prevent others from using it.

Modern American patent law states that any person who invents or discovers any new and useful process, machine, article of manufacture, or composition of matter, or any new and useful improvements thereof may obtain a patent, subject to the conditions and requirements of the law. *Process* means primarily industrial or technical processes. *Machine* and *article of manufacture* need no explanation. *Composition of matter* relates to chemical compositions, including mixtures of ingredients and new chemical compounds.

Generally, patentable inventions solve technical problems. A patentable invention must meet the Patent Office requirements. The subject matter must be *new, useful* and *unobvious to one skilled in that art. New* means the invention (as defined in the claims) was not invented by someone else before the applicant made his invention and it was not known by some member of the public more than one year before the patent application was filed. (A single copy of a thesis in the Russian language in a Chinese library has been held to be public knowledge.) *Useful* means that the subject matter is capable, at least in theory, of doing what is claimed of it. Sometimes an examiner will rquire a model to prove it works, especially when the invention hints of perpetual motion (in fact, the Patent Office will return the application and filing fee if the invention relates to perpetual motion). If the Examiner, who has an engineering degree in his field and is recognized as a world's leading authority in that area, can read the specification and study the drawings, and agree it should work with a minimum of refinements, then the invention is *useful*.

Declaring an invention *unobvious to one skilled in the art* is more difficult to decide and is very subjective. The particular Examiner may be tough or easy in his decision. If the "art" is very crowded a slight change may be "unobvious," but in a new field of technology a greater change from earlier patents, articles or publications may be "obvious" and therefore not patentable.

Sometimes a question arises as to when an inventor first "made" an invention, especially when more than one application is before the Patent Office but only one patent will be granted. To determine which inventor gets the patent, the Patent Office will determine when the project was *conceived* (when was the idea thought of) and *reduced to practice* (when first built and operated, or the patent filed). Usually the first inventor to "make" the invention gets the patent.

Surprisingly, *patent pending* has no legal meaning. It is simply a marketing technique, just another way of saying new and improved. It also is a gentle warning that someday a patent may be granted and when it is competition will be prevented from marketing the item. It is against the law and you can be fined if you don't actually have a patent application on file.

A patent gives its owner the right to exclude others from making, using, or selling the invention as defined in the claims of the country issuing the patent. An inventor can get a patent on improvements but can't make, use, or sell them if the base invention is claimed in the earlier, unexpired patent.

Making, using, or selling a patented invention without the inventor's permission is an infringement. An infringer can be enjoined from infringing or may be granted a license upon payment of royalties or other concessions.

Many inventors feel that if they could just hit upon the right idea they would get rich. Many non-patentable inventions make a lot of money while highly patentable ones may not—for example, the technology may be before its time in the commercial world. If you want to make money, invent something that great numbers of people will want and be able to buy and can be manufactured and sold cheaply. The bottle top, paper clip, hair pin, and safety match folder are prime examples. As Thomas Edison said, "There's a way to do it better—find it."

Only about half of the applications for patents ever become patents and the claimed coverages in most of them are worthless. About 10% find their way into the marketplace and about 3% make enough to pay for their cost. However, the few who follow basic rules of common sense have done very well with their brainchild. Some of the patents in this book made money and many more didn't. They all have in common the fact that they relate to something new, useful and unobvious at the time the inventions were made.

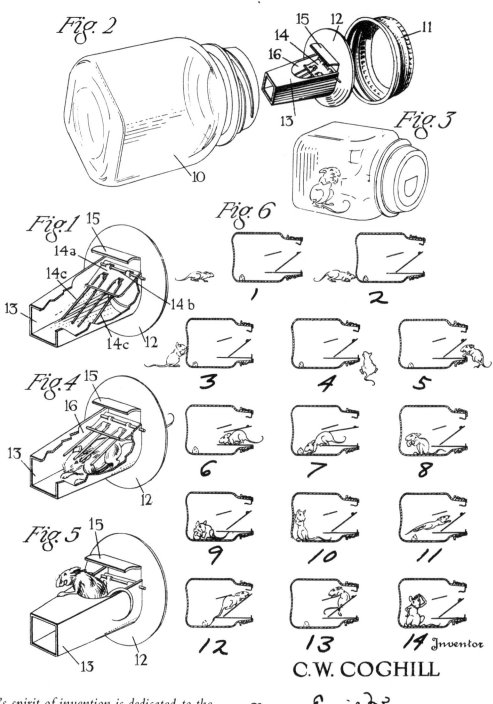

Fig. 2

Fig. 3

Fig. 1

Fig. 6

Fig. 4

Fig. 5

America's spirit of invention is dedicated to the person who can build a better mousetrap.

C.W. COGHILL

By

Inventor

Attorney

·1·
HISTORICAL PATENTS

HD
69
.N4
1787
QT
.S64

...atents started out with a strange idea. Yet ..., be they nuclear fission or the paper clip, ... a role in our lives today that it's hard to ...ut them. At the time, however, most were ...er weird or worthless.

... patent in 1790 to today, many aspects of ...d out as obscure patents. How many people ... Mestral by name? Yet try getting through the day ...t encountering Velcro. Space flight would have never gotten off the ground unless the space capsule had been patented. And proving that you needn't be rich or famous to get your name on a patent, John Stone-Parker's glass guard has earned a place in patent history—because John was all of four years old when he received it.

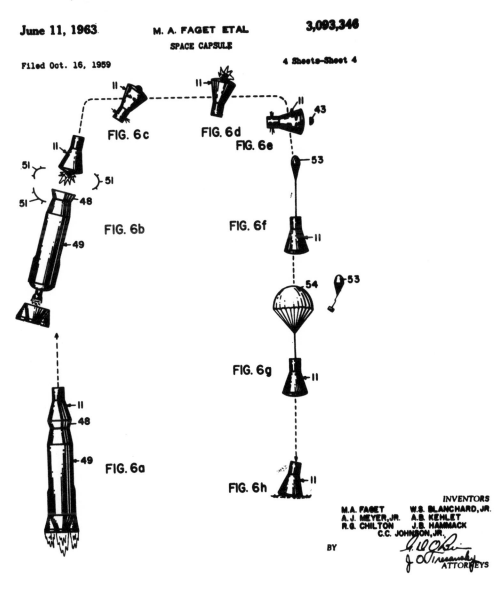

June 11, 1963.

M. A. FAGET ETAL

3,093,346

SPACE CAPSULE

Filed Oct. 16, 1959

4 Sheets-Sheet 4

FIG. 6c

FIG. 6d

FIG. 6e

FIG. 6b

FIG. 6f

FIG. 6a

FIG. 6g

FIG. 6h

INVENTORS
M.A. FAGET   W.S. BLANCHARD, JR.
A. J. MEYER, JR.   A.B. KEHLET
R.G. CHILTON   J.B. HAMMACK
C.C. JOHNSON, JR.

BY

ATTORNEYS

First Patent in America—1646

The first patent on machinery in America was granted by the state of Massachusetts in 1646 for a mill for manufacturing scythes. It consisted of a water-propelled wheel that raised a hammer which would then drop onto an anvil.

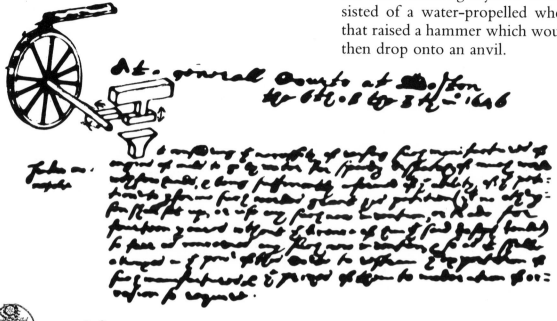

First U.S. Patent—1790

The first patent granted by the United States was for making potash, an ingredient in making soap. President George Washington signed it on July 31, 1790 and Thomas Jefferson delivered it to the inventor.

Article 1, Section 8 of the Constitution made granting patents a Federal responsibility, as a means of encouraging the disclosure of inventions.

Traction Wheels. Patent No. 1—1836

The Patent Act of July 4, 1836 established the examination system and required novelty and utility before granting a patent. Under this new patent law, Patent No. 1 was issued for traction wheels in 1836. Notched train wheels and track rails permitted a heavily loaded train to climb a steep hill without the wheels slipping.

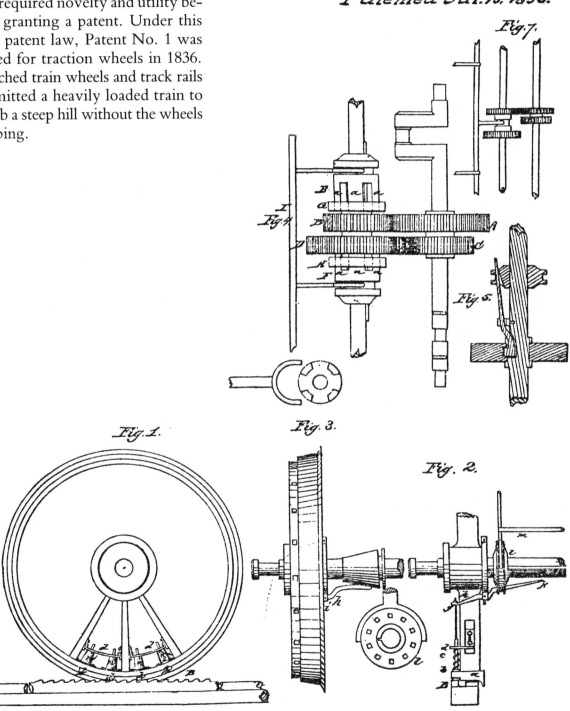

J. Ruggles.
Traction Wheels.
Patented Jul. 13, 1836.

Fig. 7.

Fig. 4.

Fig. 5.

Fig. 1.

Fig. 3.

Fig. 2.

Electric Motor. Patent
No. 132—1837

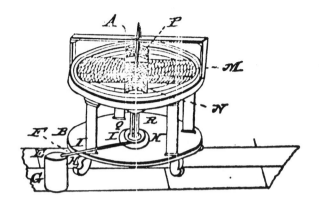

A Vermont blacksmith constructed a wheel with four electromagnets attached and connected it to a battery. The result was an electric motor.

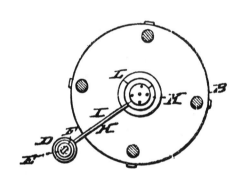

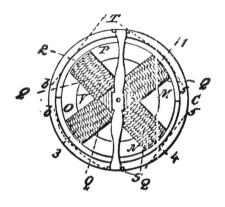

Sewing Machine. Patent
No. 4,750—1846

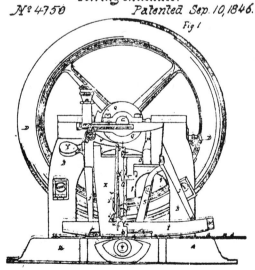

This machine provided for interlocking stitches so the thread doesn't unravel and allow the two pieces of cloth to separate.

Specifications annexed to Patent No 202 granted to Z McDaniel September 21st 1863

To all whom it may concern

Be it known that I, Z. McDaniel of Glascow in the County of Barron and State of Kentucky have invented certain new and useful improvements in Submarine Torpedoes, and I do hereby declare that the following is a full clear and exact description of

The Confederate Patent Office

Few people realize that during the Civil War the Confederate States had a Patent Office. It issued 266 patents of which about one-third related to implements of war. Rufus R. Rhodes, former examiner in the U.S. Patent Office, was its only Commissioner.

Patent No. 1 was issued on August 1, 1861 to James H. Van Houten for a Breech Loading Gun. Patent No. 100 was issued on July 29, 1862 to Lt. John M. Brooke, C.S.N. for a Ship of War. It covered many features of the ironclad rams. The 266th (and last) patent was issued on December 17, 1864 to W. N. Smith for a Percussion Cap Rammer.

The Confederate Patent Office was destroyed during the evacuation of Richmond in April 1865. The bulk, if not practically all, of the models and original records seem to have been lost. Searches in Richmond and Washington have revealed only a small trace of them. Patent 202 (above) issued Sept. 21, 1863 to Z. McDaniel for a submarine torpedo is in the permanent collection of the Eleanor S. Brockenbrough Library of the Museum of the Confederacy in Richmond, Virginia. Patent No. 15 (below) was issued August 26, 1861 to J. L. Jones for Improvement in Carriage Wheels. It provided needed reinforcement so a light spring buggy used by southern belles before the war could be used to carry supplies to the battlefront.

THE CONFEDERATE STATES OF AMERICA

To all to whom these Letters Patent shall come

Improvement in Carriage Wheels

Machine Gun. Patent No. 36,636—1862

A rapid-fire machine gun having revolving barrels. The principle of this Civil War gun is used in to-day's modern rapid-fire guns.

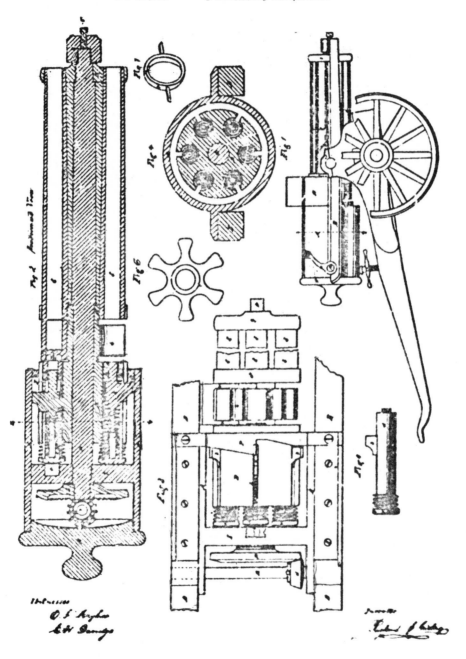

Improvement in Explosive Compounds. Reissuance No. 5,799—1874

Nitro-glycerine mixed with an absorbent substance renders the resultant explosive compound safer, more convenient for handling, storage, and transportation than liquid nitro-glycerine. Alfred Nobel, the inventor, went on to originate the Nobel Peace Prize.

Improvement in Power Car-Brakes. Patent No. 68,929—1869

The Westinghouse Air Brake made possible the safe speeding up and lengthening of trains, reducing transportation costs and enabling railroads to handle the vast traffic that became essential to modern industrial civilization.

Telegraphy. Patent No. 174,465—1876

This really was the first telephone, not telegraph. In 1887, after 2 years of hearings argued by the country's most distinguished lawyers in the cases of *McDonough v. Gray v. Bell v. Edison* to determine the original inventor of the telephone, Bell's claims were completely substantiated. Bell received 14 patents for the telephone and telegraphs, 4 for the photophone, 1 for the phonograph, 5 for aerial vehicles, 4 for hydroplanes, and 2 for the selenium cell.

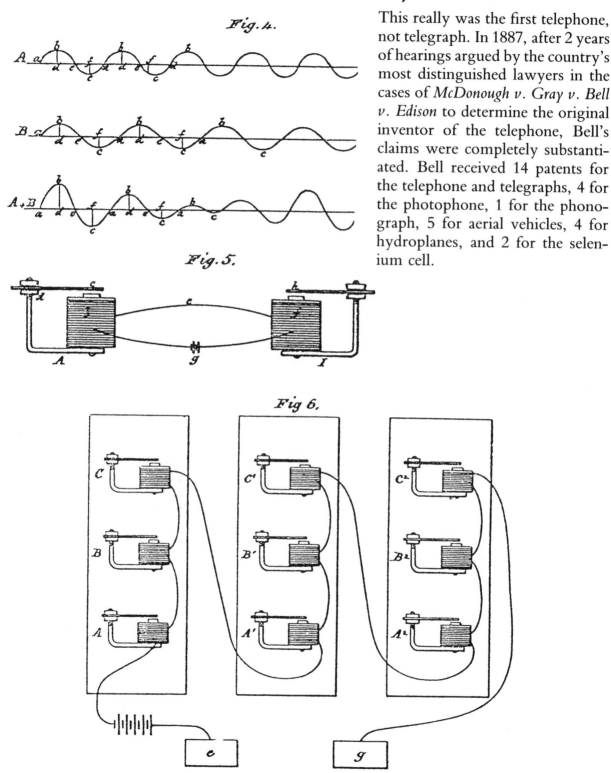

Fig. 4.

Fig. 5.

Fig 6.

Machine for Producing Type Bars & Matrices. Patent No. 345,525—1886

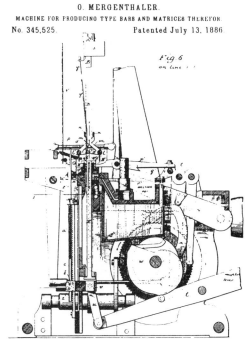

O. MERGENTHALER.
MACHINE FOR PRODUCING TYPE BARS AND MATRICES THEREFOR.
No. 345,525.   Patented July 13, 1886.

The "Linotype" machine that eliminated handset type. The operator sat at a keyboard and typed out his copy. As each key was pressed, a small brass mold dropped down onto a moving belt and was carried to an assembly box. Expansible space-bars acted to elongate or justify the line to a predetermined limit.

Camera. Patent No. 388,850—1888

After developing dry plates with his coating machine in 1880, George Eastman was able to reduce the size and weight of outdoor photographic equipment. He then developed a transparent flexible film that could be cut in narrow strips and wound onto a roller device. Film rolls for 100 exposures were mounted in a small box camera. George Eastman's roll-film camera required manual operation for all movements.

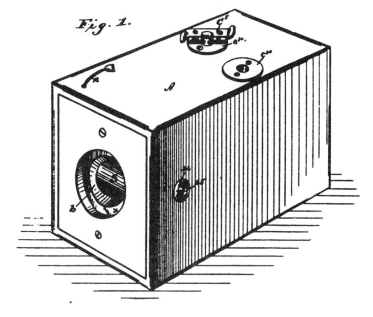

Manufacture of Aluminum.
Patent No. 400,665—1889

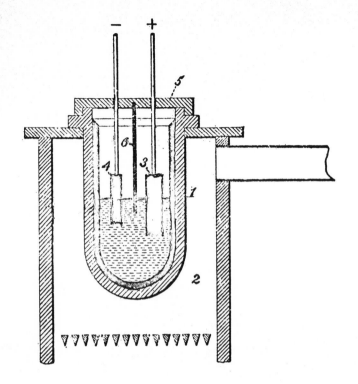

Charles M. Hall pioneered the manufacture of aluminum. This light strong metal is a rigid requirement in 22 major industries.

Internal Combustion Engine.
Patent No. 608,845—1898

Rudolf Diesel originally conceived the diesel engine as a facility adaptable in size and cost, using locally available fuels, to enable private craftsmen and artisans to compete with large industries that used the oversized, expensive, fuel-wasting steam engine as a power source. He began his 13-year ordeal creating his engine in 1885. In 1896 he demonstrated a model with a mechanical efficiency of 75.6% in contrast to the typical efficiency of the steam engine of 10% or less. Although commercial manufacture was delayed another year, by 1898 Diesel was a millionaire from franchise fees. His engines powered pipelines, electric and water plants, automobiles, trucks, marine craft, and are still used in mines, oil fields, factories and everywhere power is needed.

No. 608,845.   Patented Aug. 9, 1898.
R. DIESEL.
INTERNAL COMBUSTION ENGINE.
(Application filed July 15. 1895.)

(No Model.)   2 Sheets—Sheet 2.

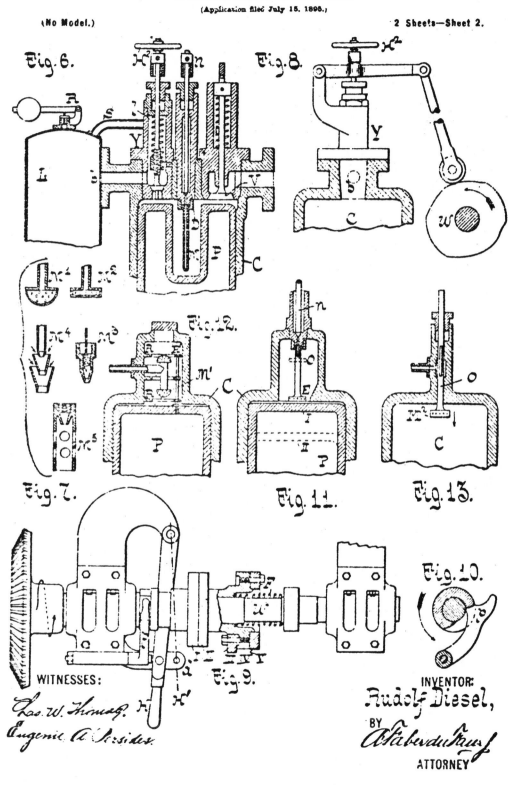

WITNESSES:

INVENTOR:
Rudolf Diesel,
BY
ATTORNEY

Spring Clip. Patent No. 742,892—1903

The first paper clip. Several versions are shown.

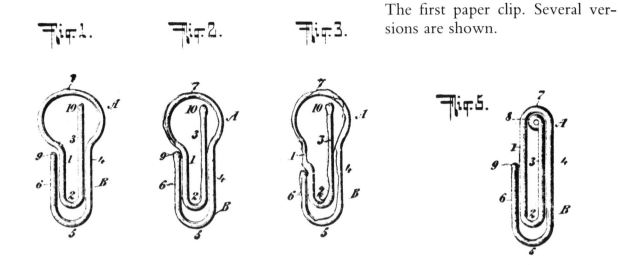

Fig.1.   Fig.2.   Fig.3.   Fig.5.

Glass Shaping Machine. Patent No. 766,768—1904

Owens glass is well known. This Owens glass shaping machine is fully automated to mold glass from a blank to a desired shape.

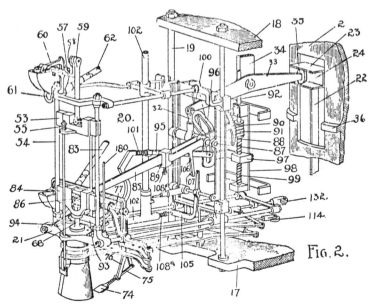

M. J. OWENS.
GLASS SHAPING MACHINE.
APPLICATION FILED APR 1?, 1903.

No. 766,768.   PATENTED AUG. 2, 1904.

FIG. 2.

Device for Amplifying Feeble Electrical Currents. Patent No. 841,387—1907

This vacuum tube was the beginning of electronics. The tube amplifies feeble electrical currents by increasing the strength of a weak signal fed into it. These weak signals were applied to a control electrode, or grid, between the cathode and anode to regulate the unidirectional flow of the current from cathode to anode. The cathode was a carbon or tungsten filament which was heated to about 2500°C by passing an electric current through it. The anode was a metal plate made positive with respect to the cathode by means of an external battery. The De Forest perforated control plate or wire grid E placed between the positive anode D and the negative cathode D^1 of a diode converted it into a triode. This control grid was held at a small negative voltage with respect to the cathode while the anode was made positive with respect to the cathode. Electrons emitted by the cathode are attracted by the positive anode, but they first have to overcome the repulsion effect of the negative grid. As a result, the electrons reaching the anode can be increased or decreased by varying the voltage on the grid. By placing the grid closer to the cathode than to the anode, a small variation in grid voltage causes a much greater variation in anode-to-cathode current. The triode can thus be used as an amplifier.

L. DE FOREST.

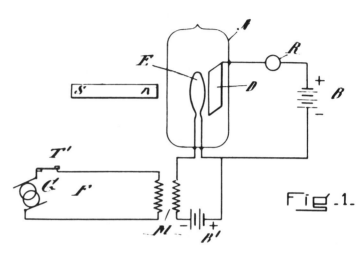

Fig.1.

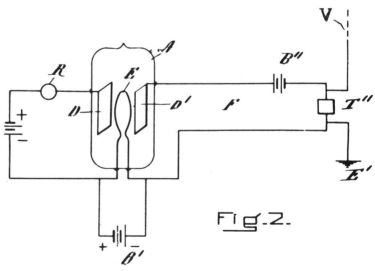

Fig.2.

Space Telegraphy. Patent No. 879,532—1908

Modern radio is born with the invention of the triode tube, having a third electrode inserted between the cathode and the anode. De Forest obtained more than 300 patents.

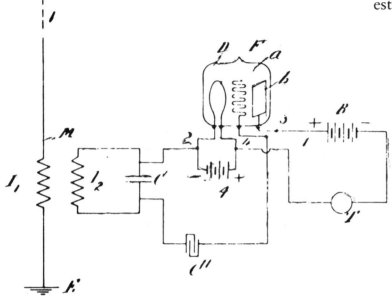

Fig. 1.

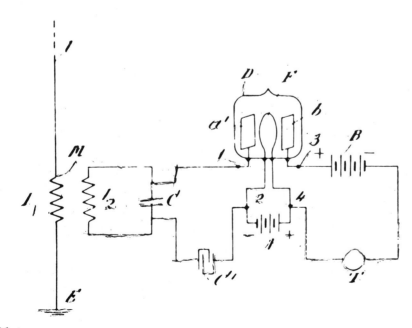

Rocket Apparatus. Patent No. 1,102,653—1914

The ancient Chinese had rockets. However, Dr. Robert Goddard's improved rocket apparatus carried photographic and other recording instruments to extreme heights. The recording apparatus does not rotate with the rocket apparatus. The elongated tapered end near the exhaust greatly increased its efficiency. After lift-off, the explosives *27* in upper chamber *25* were ignited to free the smaller upper rocket from upper firing over *24* to give the rocket a greater range.

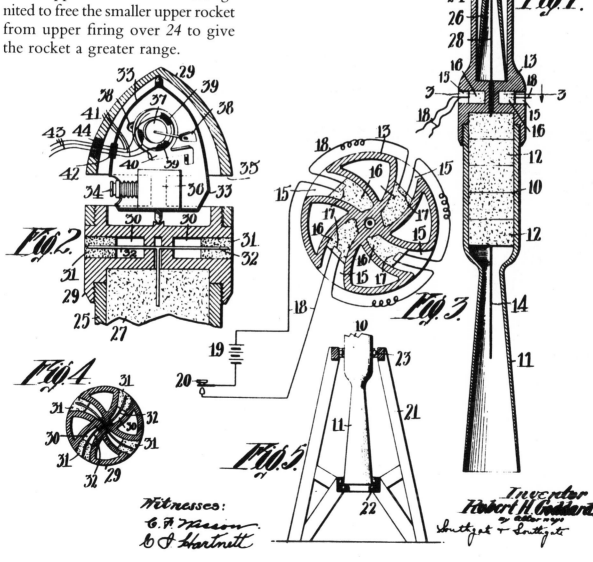

Witnesses:
C. F. Mason.
C. S. Hartnett

Inventor
Robert H. Goddard
by attorneys
Southgate & Southgate

Method of Preserving Piscatorial Products. Patent No. 1,511,824—1924

C. BIRDSEYE

METHOD OF PRESERVING PISCATORIAL PRODUCTS

Filed April 19, 1924

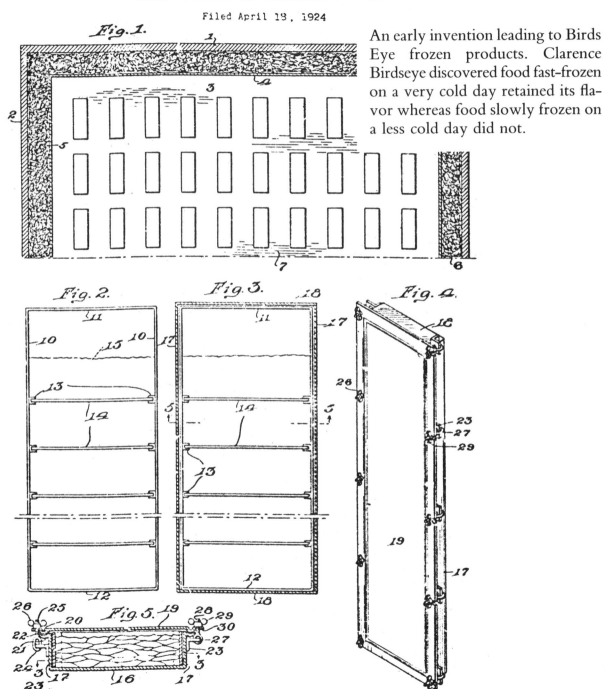

An early invention leading to Birds Eye frozen products. Clarence Birdseye discovered food fast-frozen on a very cold day retained its flavor whereas food slowly frozen on a less cold day did not.

Television System. Patent No. 1,689,847—1928

In 1884, Dr. Paul Nipkow introduced the scanning disc. This rapidly rotating metal disc with holes in it moved in front of the subject and broke up the light, dividing the picture into sections. It simply couldn't rotate fast enough to get a reasonably good picture. Dr. Vladimir Zworykin experimented with the newly discovered cathode beam because of the inconceivable speed at which it moved under the influences of electromagnets. This patent on the iconoscope scanning with an electron beam was the start of television as we know it.

V. K. ZWORYKIN

TELEVISION SYSTEM

Filed May 11, 1927   4 Sheets—Sheet 3

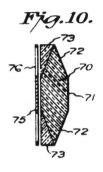

Fig.10.

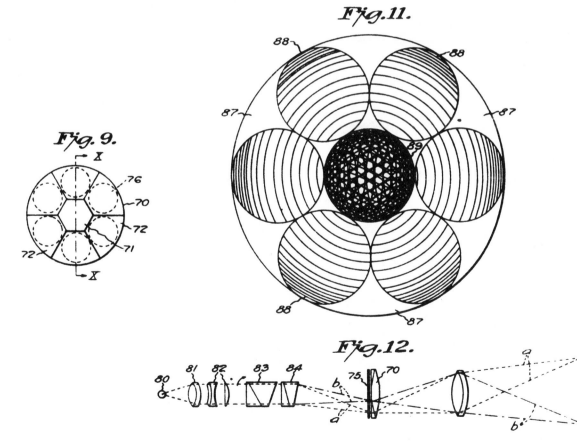

Fig.11.

Fig. 9.

Fig.12.

Neutronic Reactor. Patent No. 2,708,656—1955

It required 30 pages of specification and 27 sheets of drawings to explain Fermi's self-sustaining neutron-chain fission reactor using uranium (the first nuclear reactor). The application took 11 years before the patent was granted.

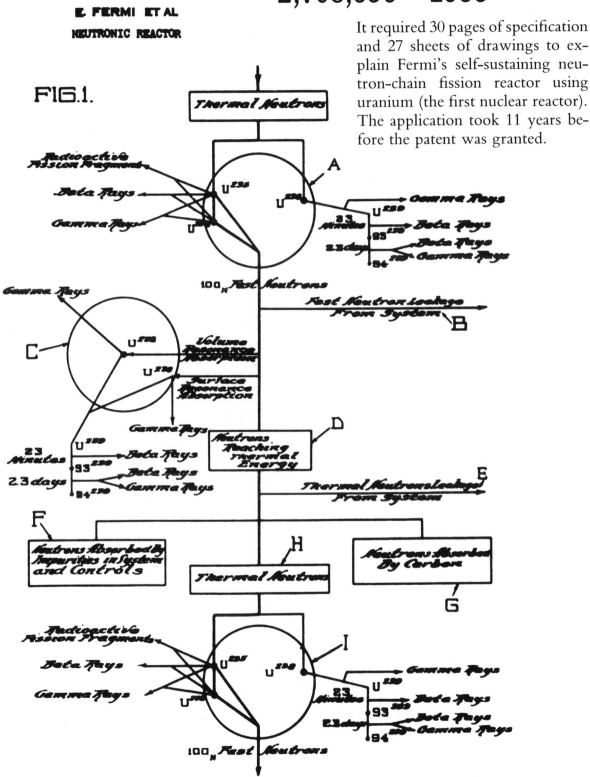

Velvet Type Fabric and Method of Producing Same. Patent No. 2,717,437—1955

A Swiss engineer was curious as to why cockleburs stuck to his socks after a walk in the woods. He examined them under a microscope and discovered hundreds of tiny hooks connected to threads of his socks. He invented a method of making the hook *4* and loop *6* configuration in nylon cloth. A bar *5* with a groove *7* is heated to give loop *6* rigidity. Knife *8* cuts the thread at groove *7* to provide thread *9* with hook *4*. The fabric can be used for a clothes or shoe brush, while two pieces may be laid over each other to adhere to each other.

Sept. 13, 1955   G. DE MESTRAL   2,717,437

VELVET TYPE FABRIC AND METHOD OF PRODUCING SAME

Filed Oct. 15, 1952

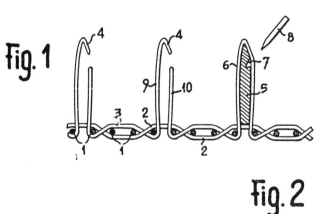

Fig. 1

Fig. 2

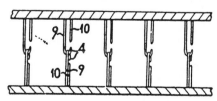

Space Capsule. Patent No. 3,093,346—1963

The beginning of human space flight. This space capsule shown in *Fig. 1* is adapted to be launched into orbital flight and returned to the Earth's surface as shown in *Figs. 6a* to *6h*.

M. A. FAGET ETAL

SPACE CAPSULE

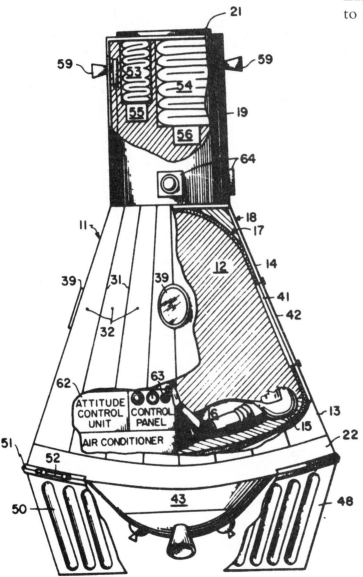

FIG. 1

Transgenic Non-Human Mammals. Patent No. 4,736,866—1988

This patent, issued April 12, 1988 to Philip Leder et al. and assigned to Harvard University for Transgenic Non-Human Mammals is the world's first patent for an animal —a genetically engineered mouse. It has caused intense controversy. It was held to be valid by the US Supreme Court ("Living matter that owes its unique existence to human intervention is patentable subject matter." *Diamond v. Chakrabarty* 1980, 477 US 303, 206 USPQ 195), which said that merely because the subject of the patent is alive and breathing does not negate an otherwise valid patent. The patent covers a live mouse injected with a substance that makes it susceptible to cancer so that the mouse could be used for laboratory experiments in cancer research.

United States Patent [19]

Leder et al.

[54] **TRANSGENIC NON-HUMAN MAMMALS**

[75] Inventors: **Philip Leder**, Chestnut Hill, Mass.; **Timothy A. Stewart**, San Francisco, Calif.

[73] Assignee: **President and Fellows of Harvard College**, Cambridge, Mass.

[21] Appl. No.: **623,774**

[22] Filed: **Jun. 22, 1984**

[51] **Int. Cl.⁴** C12N 1/00; C12Q 1/68; C12N 15/00; C12N 5/00

[52] **U.S. Cl.** **800/1**; 435/6; 435/172.3; 435/240.1; 435/240.2; 435/320; 435/317.1; 935/32; 935/59; 935/70; 935/76; 935/111

[58] **Field of Search** 435/6, 172.3, 240, 317, 435/320, 240.1, 240.2; 935/70, 76, 59, 111, 32; 800/1

[56] **References Cited**

U.S. PATENT DOCUMENTS

4,535,058 8/1985 Weinberg et al. 435/91
4,579,821 4/1986 Palmiter et al. 435/240

OTHER PUBLICATIONS

Ucker et al, Cell 27:257–266, Dec. 1981.
Ellis et al, Nature 292:506–511, Aug. 1981.
Goldfarb et al, Nature 296:404–409, Apr. 1981.
Huang et al, Cell 27:245–255, Dec. 1981.

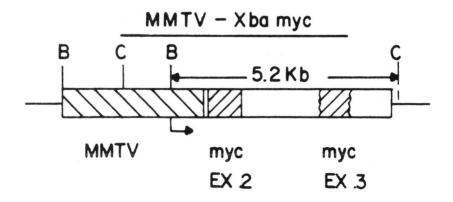

Retainer for Drinking Container. Patent No. 4,842,157—1989

This star-shaped fitting to keep ice cubes from spilling out of a glass was invented by John J. Stone-Parker with his sister Elaine. John was four years old at the time, making him the youngest patent-holder in American history. His mother, Laurie, had to read the confirmation letter to him.

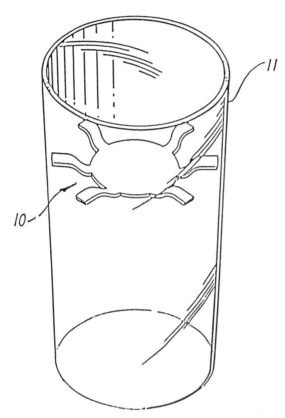

FIG. 1

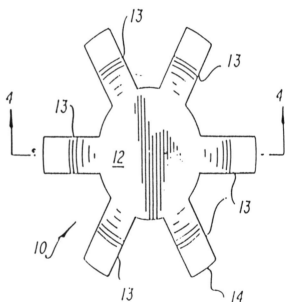

FIG. 3

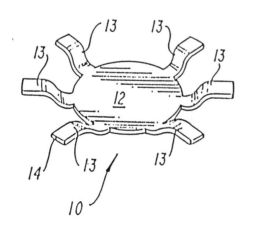

FIG. 2

·2·
FAMOUS PEOPLE/ NOT-SO-FAMOUS PATENTS

Everybody has a brainchild, even the rich and famous. Even though these people never earned anything from their filings, such luminaries as Zeppo Marx, Lillian Russell, and Walt Disney got their names on patents.

Some, like Mark Twain's game or Edgar Bergen's doll's head, are easy to understand. Others, like the secret torpedo code invented by actress Hedy Lamarr (Hedy Lamarr??) are a little more surprising. And even though he made a vast fortune in automobiles, Henry Ford's patents for a carburetor and a "motor carriage" didn't fetch him a dime.

Thomas Jefferson may have invented dozens of items before and during his presidency, but the honor of being the only president to receive a patent goes to Abraham Lincoln—for a "method of buoying vessels," so they may pass over shallow waters.

Manner of Buoying Vessels.
Patent No. 6,469—1849

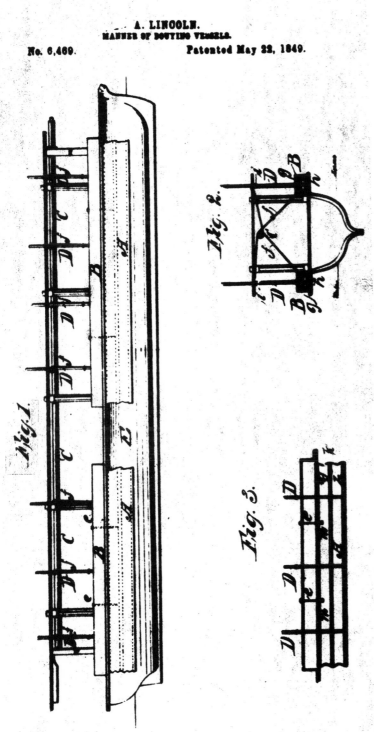

A. LINCOLN.
MANNER OF BOUYING VESSELS.

No. 6,469.   Patented May 22, 1849.

Fig. 1.

Fig. 2.

Fig. 3.

Although Abraham Lincoln was the only president to receive a patent, Thomas Jefferson is recognized as having invented the swivel chair, pedometer, shooting stick, hemp-treating machine, and an improvement in the mold board of a plow.

Lincoln's patent for buoying vessels over shoals used adjustable buoyant air chambers with a steamboat to raise the boat in the water so it could pass over bars or go through shallow water without unloading its cargo.

Game Apparatus. Patent No. 324,535—1885

Samuel Clemens (a.k.a. Mark Twain) invented this game to test the contestant's knowledge of history. Pins are placed on the player's chart in *Fig. 1* and the monarch, president, king or other ruler in that era is announced. The correct answer is on the Umpire's chart in *Fig. 2*. The game was not a success.

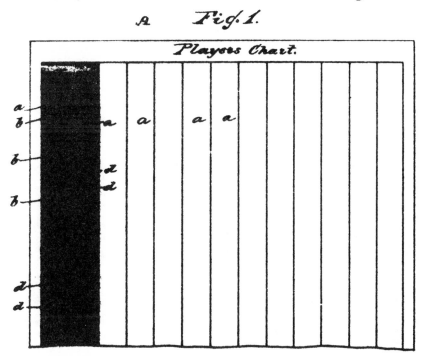

S. L. CLEMENS.

GAME APPARATUS.

No. 324,535.   Patented Aug. 18, 1885.

A   *Fig. 1.*

Carburetor. Patent No. 610,040—1898

Motor Carriage. Patent No. 688,046—1901

H. FORD.
CARBURETER.
(Application filed Apr. 7, 1897.)

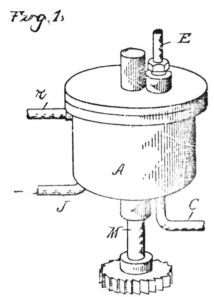

Fig. 1.

Henry Ford was granted 161 American patents (two of which are reproduced here) but was not known for his inventions. He was a businessman who believed in mass production with an assembly line of workers, each performing a small, simple task, by which in the overall, a complicated or sophisticated product was produced by many low-skilled, low-paid workers. He proudly boasted you could buy his cars in any color—as long as it was black.

H. FORD.
MOTOR CARRIAGE.
(Application filed Sept. 12, 1898.)

Fig. 8.

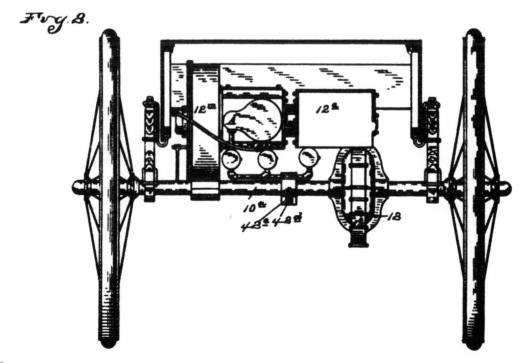

Dresser Trunk. Patent No. 1,014,853—1912

Actress Lillian Russell developed this invention. In the theatre it is a dresser. It then collapses into a trunk for travelling to the next theatre.

L. RUSSELL.
DRESSER TRUNK.
APPLICATION FILED NOV 10. 1910

Refrigeration. Patent No. 1,781,541—1927

Albert Einstein at one time was an examiner in the Swiss Patent Office. He was famous for a lot of things, but refrigeration wasn't one of them. This invention relates to apparatus and method for producing refrigeration wherein the refrigerant (butane) evaporates in the presence of an inert gas (ammonia).

A. EINSTEIN ET AL

REFRIGERATION

Filed Dec. 16 1927

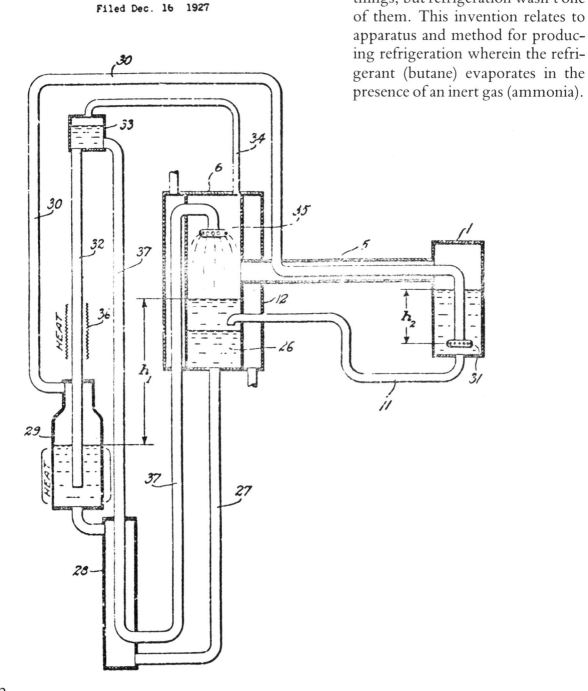

Doll Head or Similar Article.
Design Patent
No. 129,255—1941

Edgar Bergen was a star radio ventriloquist with his puppets Charlie McCarthy and Mortimer Snerd. This doll head is hand held and speaks or makes motions in response to the ventriloquist's hand inserted within the head.

E. BERGEN

DOLL HEAD OR SIMILAR ARTICLE

Filed Oct. 26, 1940

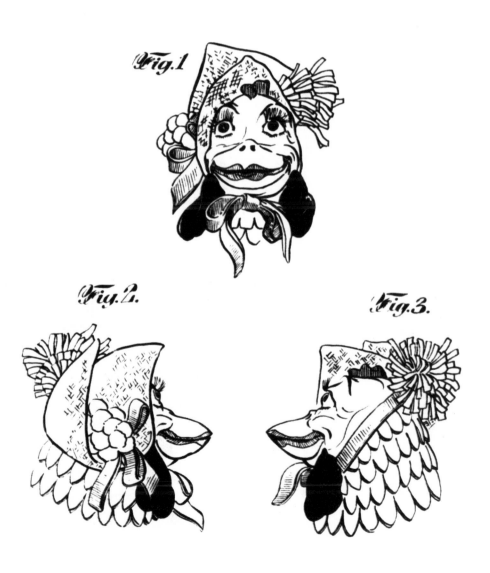

Secret Communication, System. Patent No. 2,392,387—1942

This is a secret communication system for radio control of a remote craft or torpedo. It changes the frequency of the transmitting and receiving apparatus, and uses records of the type used in player pianos—long rolls of paper with perforations. Hedy Lamarr (Hedy Kiesler Markey) was a movie actress who would not be expected by her movie-going fans to develop such a technical achievement as this.

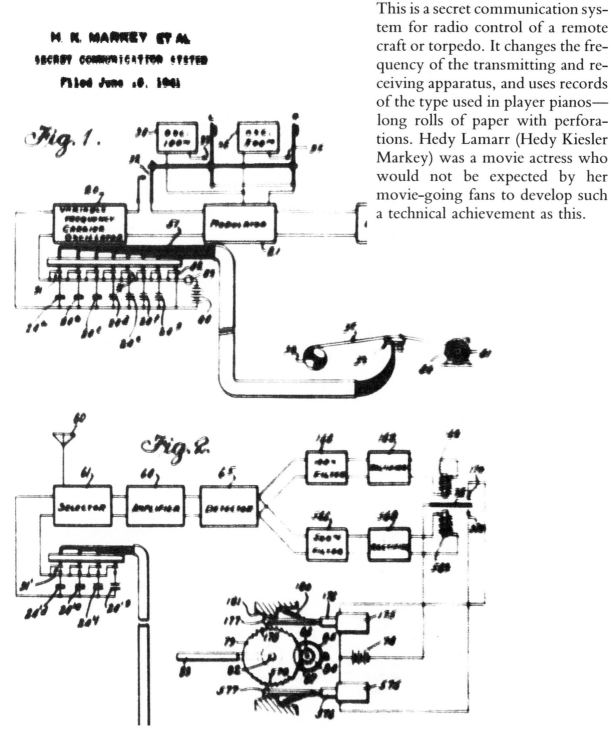

Method and Watch Mechanism for Actuation by a Cardiac Pulse. Patent No. 3,426,747—1969

Zeppo Marx of the Marx Brothers invented a double-watch wristwatch. One watch is driven by the cardiac pulse at a rate that varies according to the frequency, strength and regularity of that pulse. A second, reference, watch operates at a constant and known rate, apprising the user by reading the differences between the two watches of a corresponding variation in the functioning of the heart.

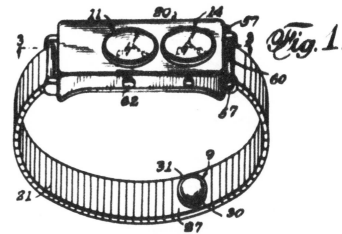

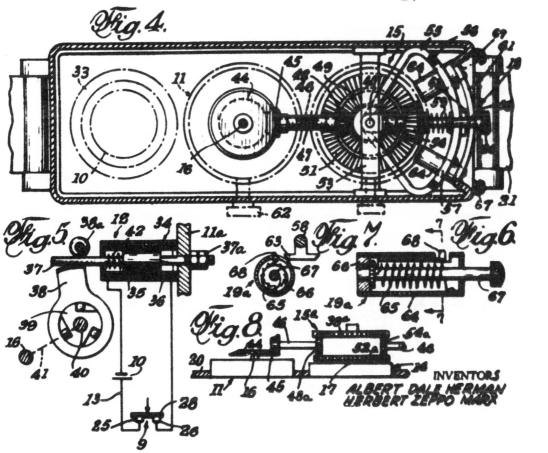

Cardiac Pulse-Rate Monitor.
Patent No. 3,473,526—1969

This monitor provides audible and vibration warning to advise persons having cardiac impairment of extreme variations in pulse rate. The unit has an electric motor operated by a small electric cell in a circuit intermittently closed by a pulse-actuated switch. An audible alarm warns of excessively high or abnormally low changes in pulse rate.

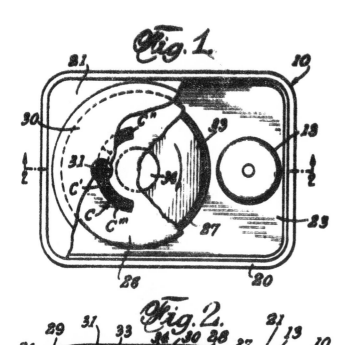

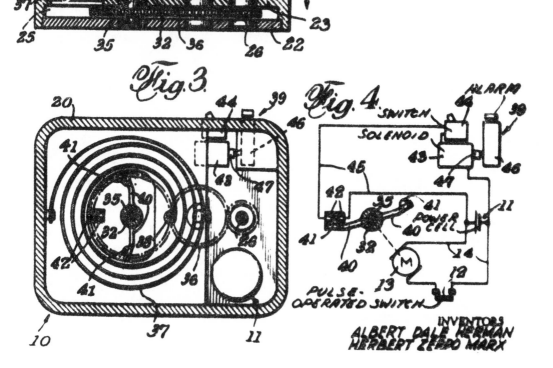

Mad Teacups. Design Patent No. 180,585—1957

Cups large enough to seat four or five children were arranged around a turntable in groups of six. The turntable revolved clockwise and each circle of six cups revolved (counterclockwise). Each cup could be entered through a hole in the side. By holding a handgrip mounted in the center like a table, a cupful of children could swing their cup around in either direction.

Patent-holder Walt Disney produced movies, built family entertainment centers, and managed a financial empire devoted to producing family fun and enjoyment.

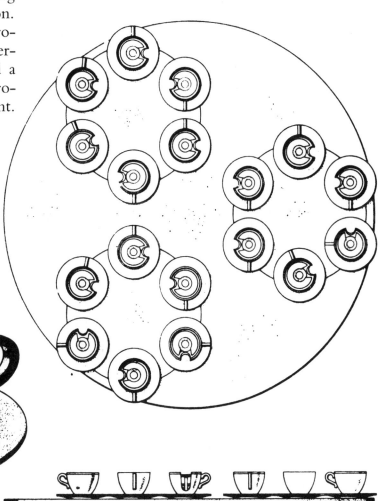

PASSENGER CARRYING AMUSEMENT DEVICE

Walter F. Disney, Los Angeles, Calif., assignor to Disneyland, Inc., Anaheim, Calif., a corporation of California

Application July 17, 1956, Serial No. 42,274

Term of patent 14 years

(Cl. D34—5)

·3·
HEALTH & BEAUTY AIDS

People are always trying to make themselves younger, thinner, tanner, prettier, stronger, more handsome and less pained. Thus, it is obviously not unlikely that many inventors would turn to this seemingly lucrative area. And for many, the rewards were plentiful.

However, for the inventors of such items as the dimple maker (which bears more than a passing resemblance to a bit brace), the cork alarm clock, or the device that allows for birth by centrifugal force, the market was somewhat lacking. Yet it remains a tribute to ingenuity that someone who was dissatisfied with the shape of their (or their spouse's) upper lip could produce a device which would alter it to its desired appearance—a true alternative to lipstick.

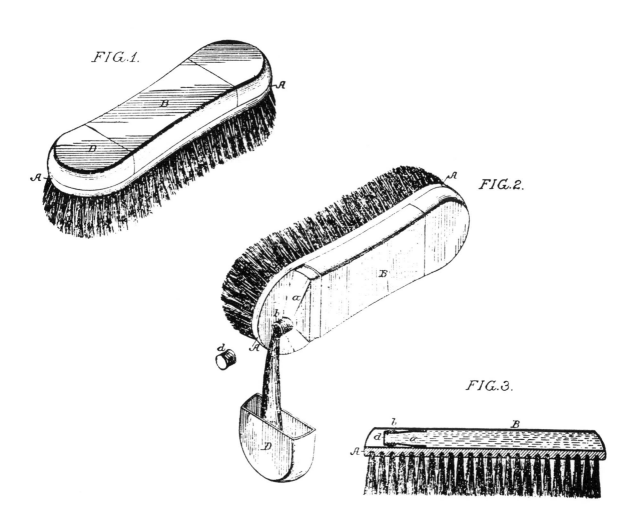

Napkin Holder. Patent No. 541,384—1895

B. F. PASCOE.
NAPKIN HOLDER.

This napkin holder has two attached pieces of metal. One end forms a pair of tweezers to grip the edge of an open napkin. The other end is a hook to fit in the collar. It also can be used as a napkin ring to hold a folded napkin together.

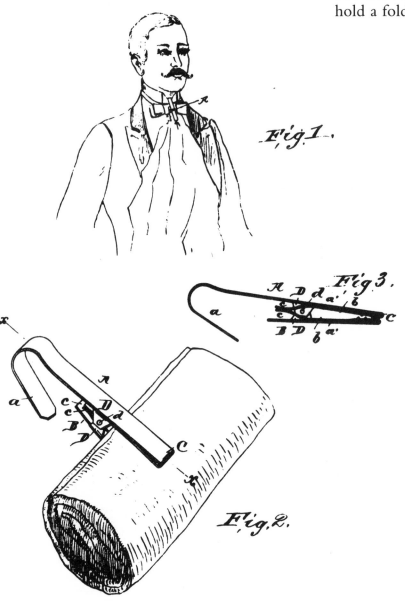

Fig.1.

Fig.3.

Fig.2.

Device for Preventing Snoring.
Patent No. 587,350—1897

Snoring is caused by breathing with the mouth open. This device prevents it by causing the volume of air to be broken as it passes through slots in the central portion of the mouthpiece. An elastic band keeps the slotted mouthpiece in place.

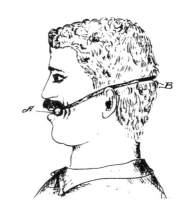

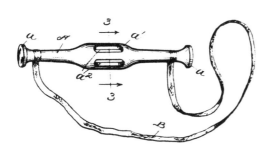

Electrical Bedbug Extermination. Patent No. 616,049—1898

This bedbug exterminator has electrical devices, which are applied to the bedstead. This allows currents of electricity to go through the bodies of the bugs, which will either kill them or startle them into leaving. Pairs of insulated contact-strips are used to hold the bug's body to complete the circuit, thus exterminating the bug.

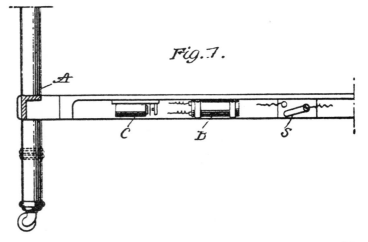

ELECTRICAL BEDBUG EXTERMINATOR.
(Application filed Feb. 7, 1898.)

Fig. 1.

53

Device for Inducing Sleep.
Patent No. 313,516—1885

F. W. PAUL.
DEVICE FOR INDUCING SLEEP.

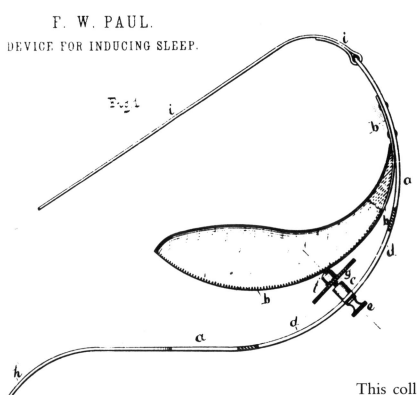

Fig. 1

This collar applies pressure to the arteries and veins of the neck. A padded spring inside the band of the collar presses against the flesh below the jaw and midway between the ear and chin. A thumb-screw adapts the device to necks of different sizes. After the undesirable flow of blood is controlled, the nervous system becomes soothed and sleep follows almost immediately.

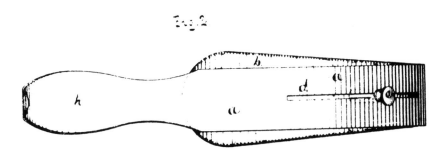

Fig. 2

Moustache Guard. Patent No. 411,988—1889

A moustache guard attachment for spoons or cups. When eating soup or drinking, the spoon or cup is raised to the mouth. Guard *B* passes under the moustache and raises it so it will not touch the contents of the spoon or cup.

M. E. HARRINGTON.
MUSTACHE GUARD.

Patented Oct. 1, 1889.

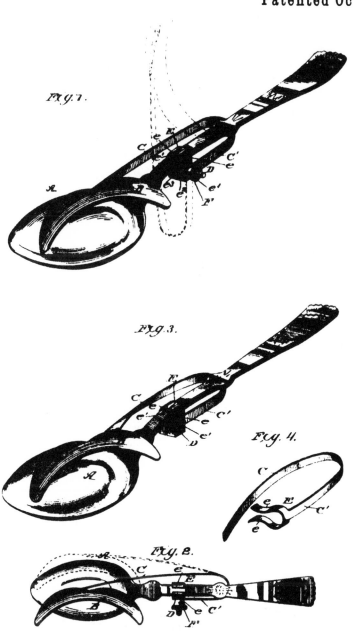

Foot Warmer. 1815

Hot ambers from a fireplace are shoveled into the foot warmer and the door clicked shut. The ambers stay hot for many hours and the metal box prevents them from starting an accidental fire.

A. Greely.

Foot Warmer.

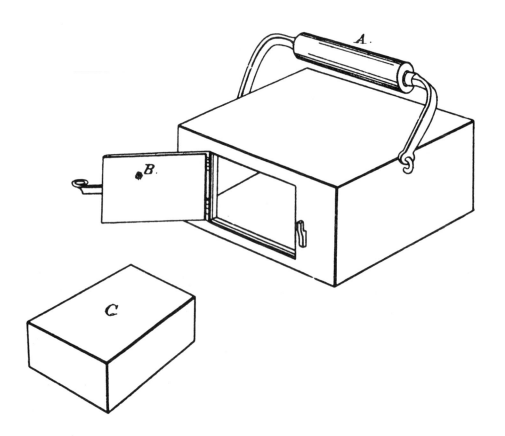

Moustache Guard. Patent No. 680,578—1901

This moustache guard holds the moustache away from the lips to prevent food from lodging in it. It may be carried in the vest pocket. The user inserts a series of up-wardly-inclined teeth and then straps an elastic strip around the moustache. Prongs on the guard support the long flowing ends of the moustache to keep them from drooping.

Mirror Attachment for Table Implements. Patent No. 886,746—1908

Knives and forks with reflectors or mirrors set in their handles to aid the diner in dislodging food from between his teeth or to see if his lips are clean.

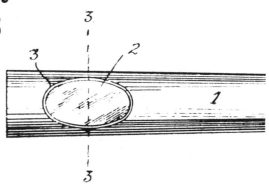

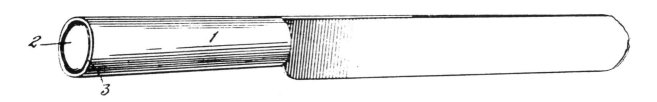

Combined Clothes Brush, Flask, and Drinking Cup.
Patent No. 490,964—1893

Fig. 1 shows what appears to be an ordinary clothes brush. Cup *D*, however, may be readily removed and used for drinking purposes as shown in *Fig. 2*.

T. W. HELM
COMBINED CLOTHES BRUSH, FLASK, AND DRINKING CUP.
No. 490,964.   Patented Jan. 31, 1893.

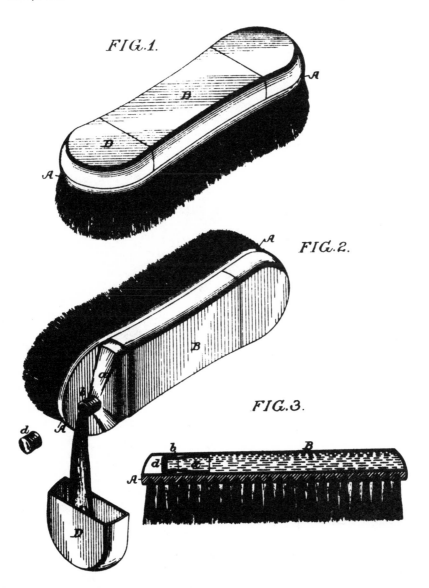

Device for Producing Dimples.
Patent No. 560,351—1896

Knob *C* is placed where the dimple is to be formed. Knob *N* is held by one hand and brace *I* is revolved in the same manner as an ordinary bit brace tool. Cylinder *F* causes skin to mass and make the skin malleable around the spot where the dimple is to be produced.

DEVICE FOR PRODUCING DIMPLES.

Patented May 19, 1896.

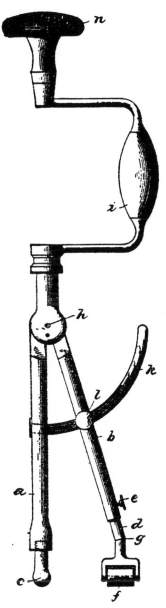

Device for Shaping the Upper Lip. Patent No. 1,407,342—1924

This device reshapes the upper lip of a person to confirm to a "Cupid's bow." It forms a depression in the upper surface and draws the upper lip into a shaping relationship with the matrix.

DEVICE FOR SHAPING THE UPPER LIP

Filed March 25, 1922

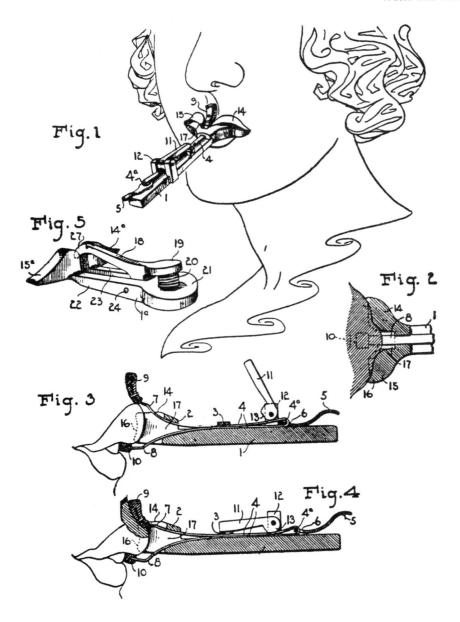

Method for Minimizing Facial Aging. Patent No. 2,619,084—1952

A series of little anchors is inserted in the scalp and joined by rubber bands. This draws the skin taut over the face and holds it for several hours. Wrinkles temporarily disappear, providing a youthful appearance "until the party is over."

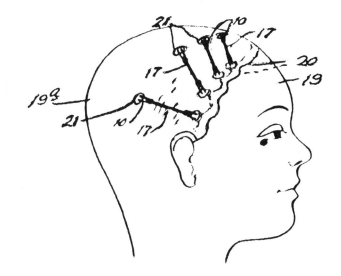

Shaving Gauge. Patent No. 1,504,436—1924

This gauge over the ear accurately limits shaving below the sideburns.

D. DAUGHLEY

SHAVING GAUGE

Filed Oct. 16, 1923

61

Spinnet. Patent No. 3,216,423—1965

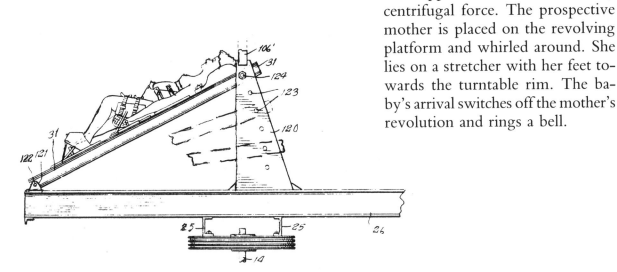

This apparatus aids childbirth by centrifugal force. The prospective mother is placed on the revolving platform and whirled around. She lies on a stretcher with her feet towards the turntable rim. The baby's arrival switches off the mother's revolution and rings a bell.

Smoking Deterrent. Patent No. 3,655,325—1972

This cigarette package starts coughing loudly when someone picks it up. It is intended to help cure the habit and also as an advertising novelty.

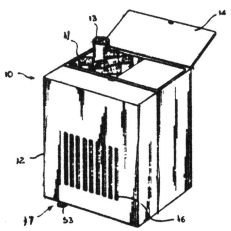

·4·
HELP AROUND
THE HOUSE

Just as people are always trying to improve themselves, they are always trying to improve their habitats. Before the latter half of the 20th century ushered in the age of high-technology household gadgets, housewives (and their husbands) were pretty much left to their own devices. Even still, they were able to come up with a multitude of step-saving items (prominent among them the combination writing desk and toilet-box).

The ability to perform two tasks at once was a common theme among these inventions—a bath tub that allowed one to bathe while swiveling to avoid bathtub ring, a swing that churned butter and ran a washing machine, and a book which hid a liquor flask (apparently for when the work was done).

Pencil & Eraser. Patent No. 19,783—1858

A pencil with an eraser on the upper end. Later the courts held that the "combination" of a lead marker on one end and an eraser on the other was really an "aggregation," which is not validly patentable.

H. L. Lipman.
Pencil & Eraser.
N°. 19,783. Patented Mar. 30, 1858

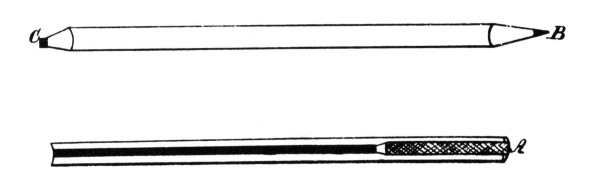

Sewing Machine. Patent No. 22,148—1858

This is a mechanical improvement patent on the looper, feeder and tension portions of a sewing machine—but nothing is said about the horse. Perhaps it was added to draw attention to the patent and further promote the invention.

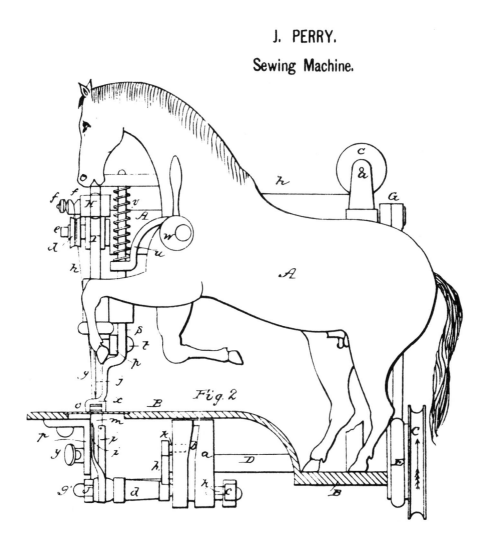

J. PERRY.

Sewing Machine.

Fig. 2

Wardrobe, Piano, and Bedstead. Patent No. 56,413—1866

This space saver combines a bed *J*, sofa *I*, piano *A*, chest of drawers *E*, and wardrobe *C*. However, bulkiness presented a problem for movers climbing stairs and moving through doorways.

C. Hess,

Wardrobe, Piano and Bedstead

Toilet, Work-Box and Writing-Desk. Patent No. 211,656—1879

This combination toilet case, workbox and writing desk has a mountable back *A*, to which toilet case *B* is attached. It holds brushes, combs, cosmetics, and other required toilet articles. Hinged to the toilet case *B* is a swinging portion *C*, one side of which is a workbox *F*, including table *b*. On the other side of *C* is a desk *D*. Hinged to *D* is a lid *E*, which drops down to form a leaf upon which to write.

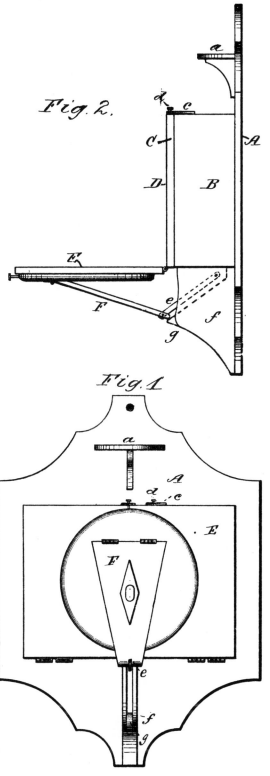

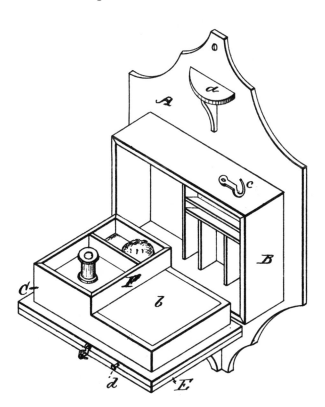

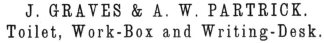

J. GRAVES & A. W. PARTRICK.
Toilet, Work-Box and Writing-Desk.

Combined Step-Ladder and Ironing-Table. Patent No. 215,171—1879

H. C. & E. L. SHANAHAN.
Combined Step-Ladder and Ironing-Table.

No. 215,171.   Patented May 6, 1879.

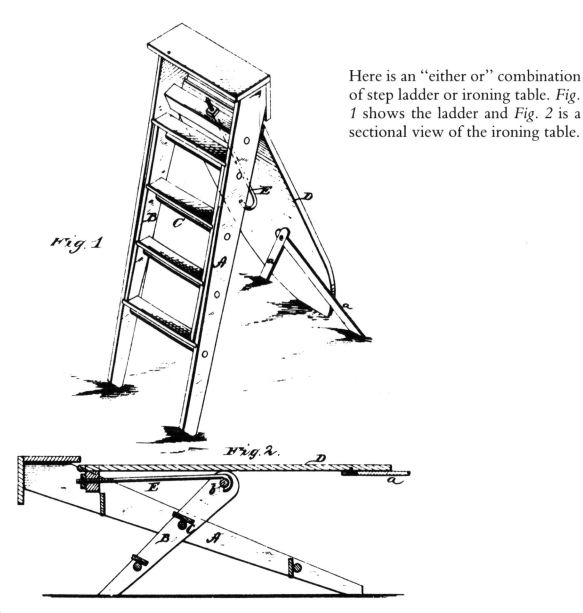

Here is an "either or" combination of step ladder or ironing table. *Fig. 1* shows the ladder and *Fig. 2* is a sectional view of the ironing table.

Device for Waking Persons from Sleep. Patent No. 256,265—1882

Instead of bells or audible alarms, a frame is suspended over the sleeper and attached to a clock. When the set time occurs, the frame is lowered to the sleeper's face. The dangling corks strike the sleeper gently to awaken him without causing pain.

S. S. APPLEGATE.
DEVICE FOR WAKING PERSONS FROM SLEEP.

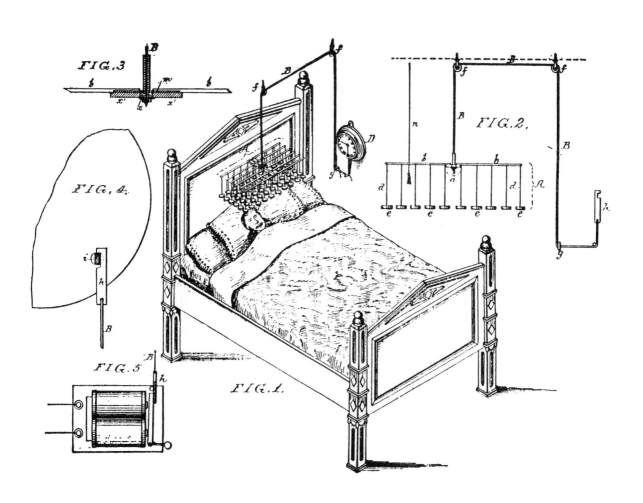

Child's Carriage. Patent No. 308,467—1884

This baby carriage was designed as a very fancy high-top shoe complete with laces and an umbrella to keep rain or sun off the baby's head.

Device for Operating Churns, Washing Machines, etc. Patent No. 383,010—1888

J. RESTEIN.

DEVICE FOR OPERATING CHURNS, WASHING MACHINES, &c.

As the girl swings, lever *K* moves back and forth. This operates the dasher of the appliance *L* which may be a washer, butter churn or similar device.

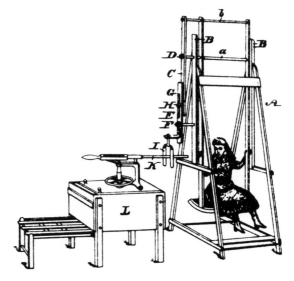

Combined Match Safe, Pincushion and Trap. Patent No. 439,467—1890

Cushion *B* and match-tray *C* are removed from main box *A*. The box bottom is opened downward as shown in *Fig. 4*. Baited hook *G* is set to hold the bottom open until a mouse trips it, causing the box to fall and capture the mouse. Immersion in water drowns the mouse.

H. BRANDT.
COMBINED MATCH SAFE, PINCUSHION, AND TRAP.

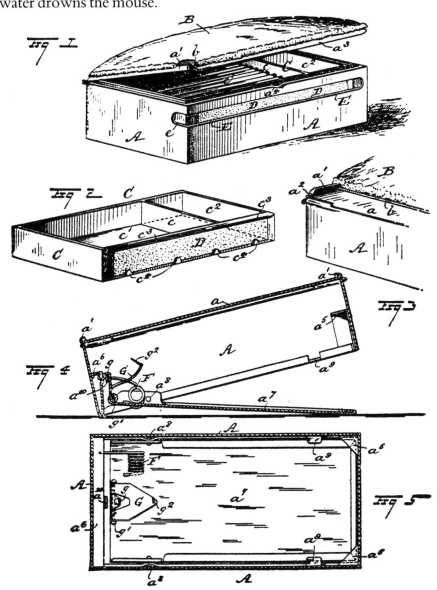

Liquor Flask. Patent No. 330,709—1885

A container *A* fits within the book *B*. An opening *D* in the bottom permits insertion of a finger to push the container *A* upwards. The neck *E* of the container is pushed up through doors *e* to provide access to the cap *a*. After the cap *a* is replaced, the container *A* drops down, and the flask again resembles a book.

H. W. T. JENNER.

LIQUOR FLASK.

No. 330,709.

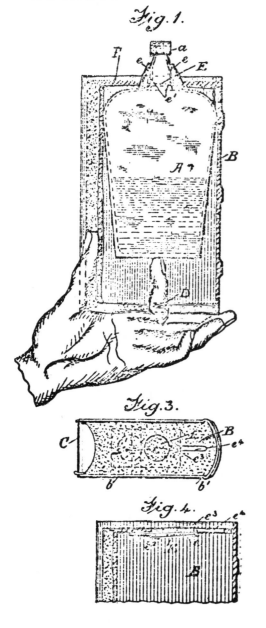

Fig.1.

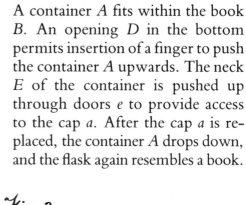

Fig.2.

LEGAL DECISIONS

VOL. II.

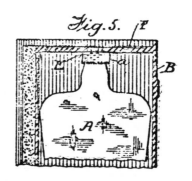

Fig.3.

Fig.4.

Fig.5.

Rocking or Oscillating Bath Tub. Patent No. 643,094—1900

This rocking bathtub prevents tub ring and is supposed to have therapeutic value to the bather.

O. A. HENSEL.
ROCKING OR OSCILLATING BATH TUB.
(Application filed Jan. 6, 1899.)

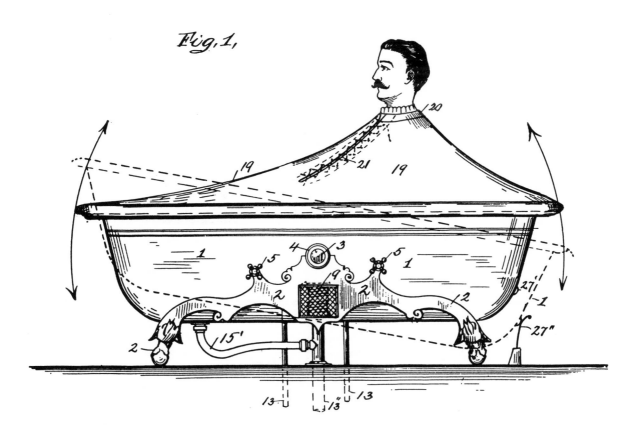

Churn. Patent No. 1,051,684—1913

This churn is rotated by a person seated and rocking in a rocking chair.

CHURN.

APPLICATION FILED JUNE 28, 1912.

1,051,684.

Patented Jan. 28, 1913.

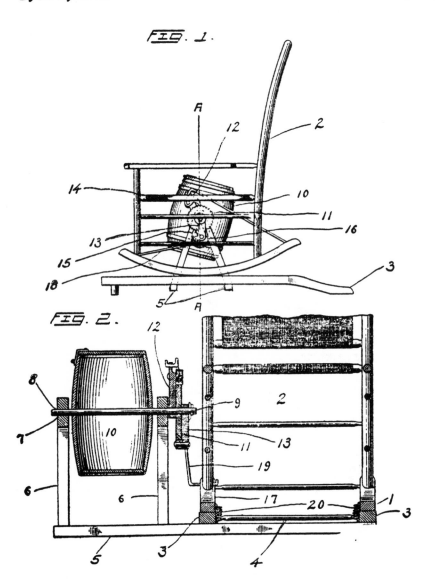

FIG. 1.

FIG. 2.

Protective Means for the Containers of Valuables. Patent No. 1,563,176—1925

Tearing the bag from the carrier's hand releases a belching cloud of white smoke caused by exposure to the air of titanium tetrachloride. Tear gas can also be released from another container.

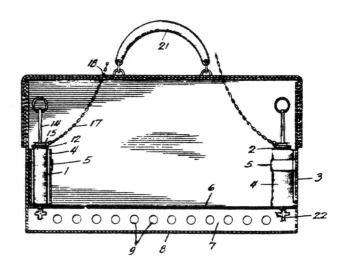

Mouth Opening Alarm. Patent No. 2,999,232—1961

When the wearer's jaw drops during sleep, a battery-powered vibrator and audible alarm is set off. This eventually conditions the wearer to sleep with his mouth closed.

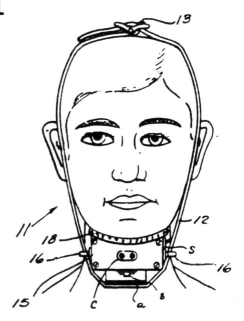

Suction Cleaning Apparatus.
Patent No. 1,105,942—1914

Foot operated bellows *1* and *2* strapped onto the feet of the operator create a suction at nozzle *22* through hoses *17*, *18* and *24*. When the operator bears weight on bellows *1*, valve *6* exhausts the air. When the operator bears weight on bellows *2*, bellows *1* enlarges under springs *9* to create the vacuum which extends to the nozzle *22*. By jogging in place, furniture, curtains, and walls can be vacuumed also.

E. M. WARING.
SUCTION CLEANING APPARATUS.
APPLICATION FILED MAR. 16, 1909.

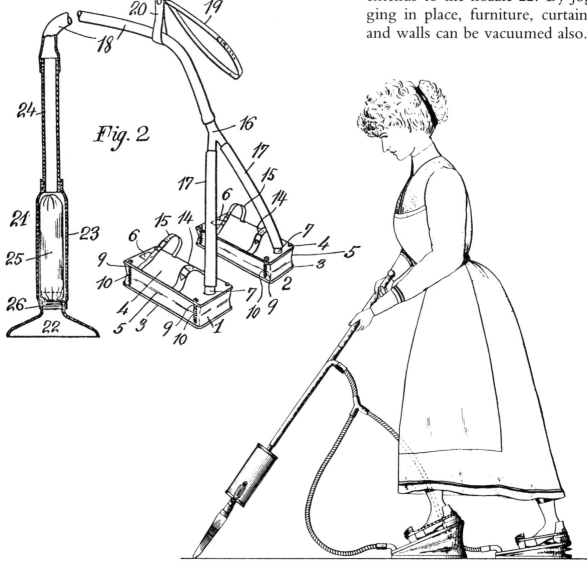

Fig. 2

·5·
THE FARMING LIFE

Farming is the backbone of America, so it is no wonder that many of the patents granted by the United States Patent Office have to do with agriculture.

Most of these inventions were devised to fit a specific need. For example, the combined plow and shotgun would be better served for hostile areas, the sanitary cow stall would not be much use for chicken farmers (vice versa for the hen blinders), and artificial pig suckling devices would have only limited appeal.

It is interesting to note that most of these inventions were put together from items that would be commonly found around the farm, mostly because they were devised out of the actual needs of the moment (see the portable milking stool and cow mask).

Combined Plow and Gun.
Patent No. 35,600—1862

This combined gun and plow was useful in border areas, which were subject to guerilla warfare and Indian raids. A projectile of one to three pounds did not render it cumbersome as a plow.

COMBINED PLOW AND GUN.

Patented June 17, 1862

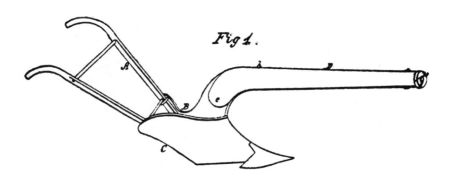

Fig 1.

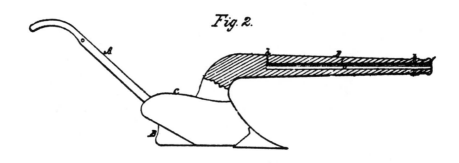

Fig. 2.

Foot Corn Planter. Patent No. 37,922—1863

A grain pocket is connected to the farmer's shoe through a tube. As the farmer walks, the ground is pierced and seed falls into the hole.

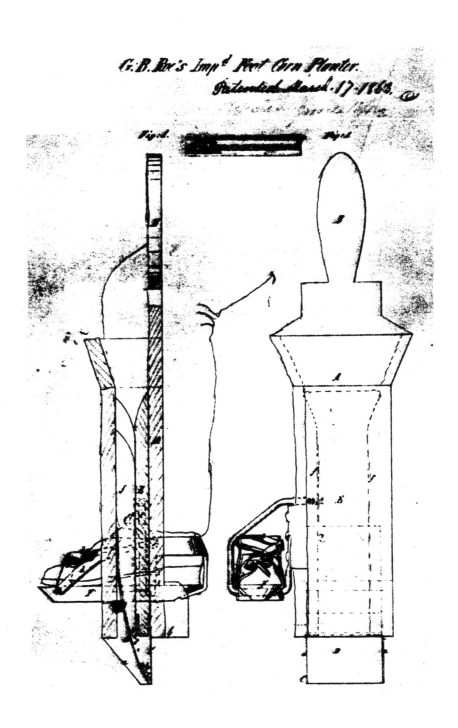

Horseshoe. Patent No. 44,603—1864

M. Chittenden,

Horseshoe.

N⁰ 44,603.

These interchangeable horseshoes are easily removable and can be repaired at the owner's leisure. Spare sets may be used if necessary. A smooth shoe, one with calks, may be greatly conducive to the health and vigor of the animal, according to the application.

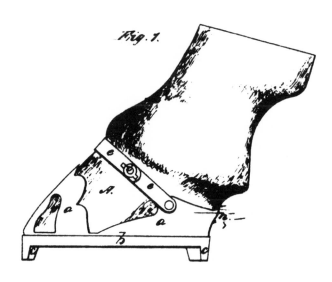

Fig. 1.

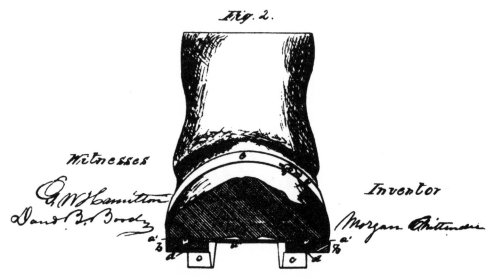

Fig. 2.

Witnesses

G. W. Hamilton

David B. Boody

Inventor

Morgan Chittenden

Method of Precipitating Rainfall. Patent No. 230,067—1880

A method of producing rainfall by exploding blasting agents in cloud areas. Explosives are transported by balloon and detonated by electric shocks through a wire running from a battery on the ground.

D. RUGGLES.
Method of Precipitating Rain Falls.

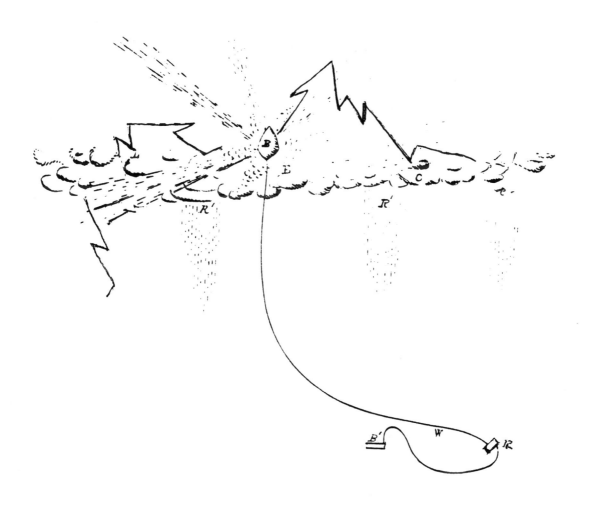

Milking Stool. Patent No. 359,921—1887

This milking stool straps around the waist of the milker and dangles behind. When ready to milk, the wearer leans forward and the stool swings beneath the milker without being touched by either hand. If the cow moves, the milker can get up and look after the milk buckets without paying attention to the stool.

A. B. COWAN.

MILKING STOOL.

Fig. 1.

Fig. 2.

Fig. 3.

Device for Preventing Hens from Setting. Patent No. 582,320—1897

When a hen is provided with this hood, she cannot see right, left, or upwards. All nests in modern henneries are at an elevation above ground and since a fowl will never fly where it cannot see, it will not fly up to the nest. It will also keep the bird from flying over a fence and into gardens.

A. J. SPARKS.
DEVICE FOR PREVENTING HENS FROM SETTING.

Eye Protector for Chickens.
Patent No. 730,918—1903

This eye protector for fowls protects them from other fowls that might attempt to peck them. It is easily and quickly applied and does not interfere with the fowl's eyesight.

EYE PROTECTOR FOR CHICKENS.
APPLICATION FILED DEC. 10, 1902.

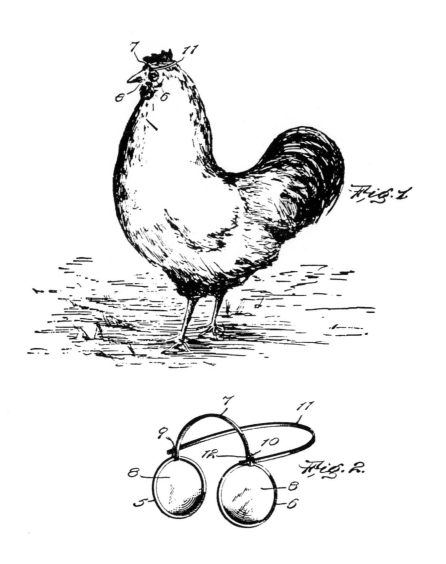

Feeding Device for Poultry.
Patent No. 828,227—1906

The inclined rotatable platform *9* in front of feedbox *13* continuously moves a chicken out of range as it tries to feed from the box. To stay within range the chicken must walk rapidly and thus it gets the desired exercise.

FEEDING DEVICE FOR POULTRY.
APPLICATION FILED MAR. 21. 1906.

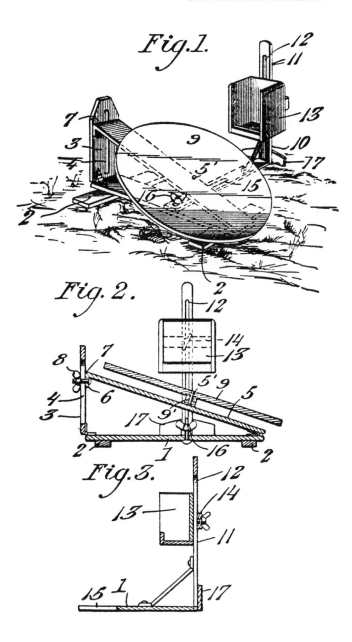

Apparatus for Irrigating Plants. Patent No. 1,278,217—1918

This machine makes and fires ice pellets *33* into the ground where they slowly melt. This is especially beneficial in climates of high temperatures and dry atmospheres where surface irrigation is not practical.

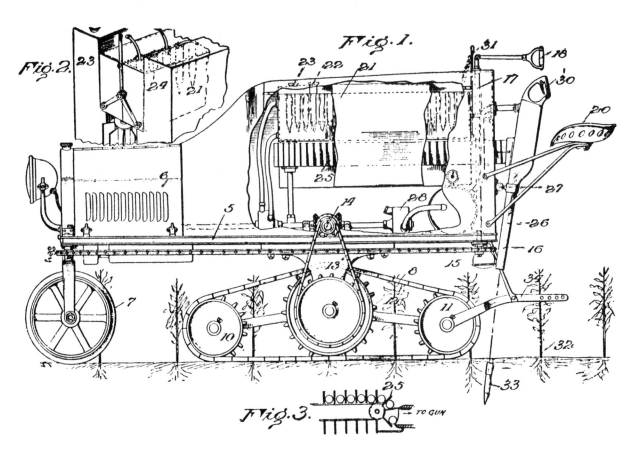

APPARATUS FOR IRRIGATING PLANTS.
APPLICATION FILED JUNE 8, 1918.

1,278,217.

Patented Sept. 10, 1918.

Milker's Mask. Patent No. 1,290,140—1919

A catcher's mask for a cow milker. The wiring across the face is much closer than the bars on a catcher's mask because flies are smaller than baseballs.

Poultry Disinfector. Patent No. 1,303,851—1919

This disinfector sprays insecticide powder on poultry. Weight on the platform *12* causes the piston *27* to rise, forcing air through the pipe *24* to the nozzle *16*. Air under pressure stirs the powder in the hopper *21* to pass some into the pipe and onto the chicken.

POULTRY DISINFECTOR.
APPLICATION FILED SEPT. 11, 1918.

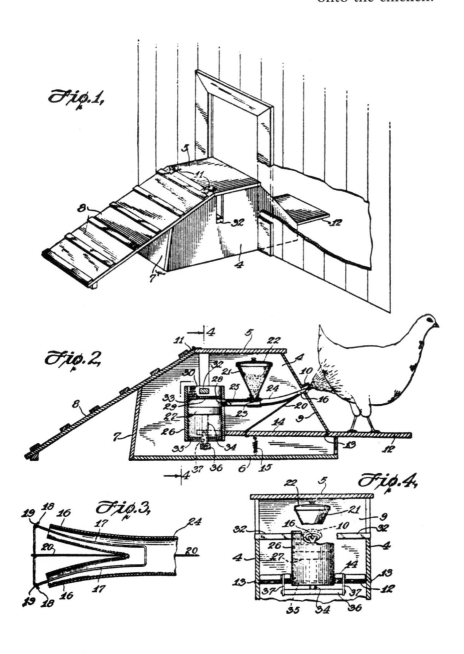

Sanitary Cow Stall. Patent No. 1,512,610—1924

This frame has a row of pins *12* on a bar *15*. This bar is adjustably mounted over the cow's back. When the cow humps up prior to going to the bathroom, the pins prick its back and the pain discourages the cow's action until it backs out of its stall. This keeps the stall sanitary.

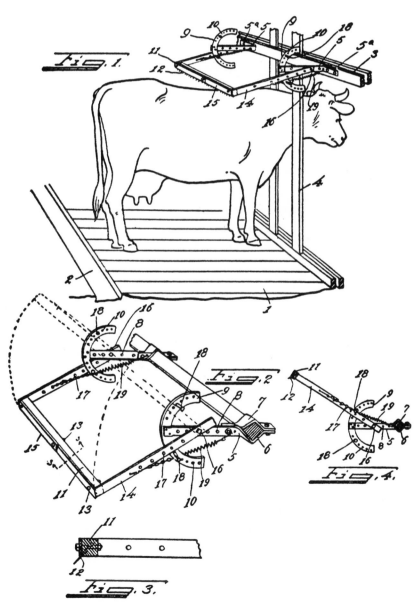

SANITARY COW STALL

Filed March 22 , 1924

Face Fly Mask for Dairy Cows. Patent No. 3,104,508—1963

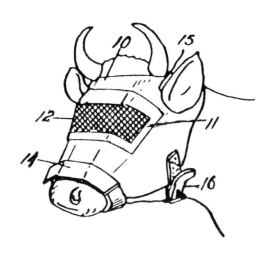

A plastic shield *10* is placed over the cow's face and held by a strap under the jaw. A screen *12* lets the cow see. It is not unpleasant for the cow and ensures continued milk production.

Artificial Pig Feeder Suckling. Patent No. 3,122,130—1964

The mechanical mother has two rows of nipples. A timed operating cycle pumps warm milk while a tape plays through a speaker.

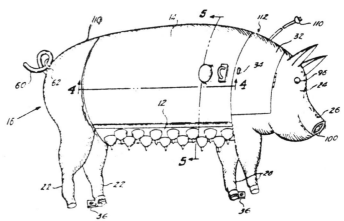

·6·
TRANSPORTATION

Locomotion has always been a vital concern, ever since feet first got tired after a long trip. So obsessed were inventors in the 19th century with personal transportation that the USPO received hundreds of patents for a *velocipede*—technically a bicycle, but coming to mean any self-propelled vehicle. Dogs, horses, birds, and even humans were used to power them. While quaint by today's standards, to produce an effective and marketable velocipede was the equivalent of finding the end of the rainbow to many inventors.

Other transportation inventions came, again, from sheer (or perceived) need. Improvements in safety and comfort were always in demand. To save their precious cattle from being killed by interloping trains (or save their precious trains for being damaged by stubborn animals), many patents were filed for pushing livestock off railroad tracks. Likewise, to keep loved ones from being killed by interloping cars (or keep cars from being damaged by slow people), inventors sought to outfit early autos with similar devices. And, since many believe the oldest solutions are often the best solutions, the USPO issued Patent No. 1,469,110—for a message in a bottle.

Velocipede—1804

The gearing is such that the man at
the rear can rotate the crank to move
the vehicle forward while the man
at the front can steer.

J. BOLTON.

Velocipede.

Velocipede. Patent No. 223,241—1880

A durable velocipede to be used by either boys or girls. Shields project over the treads of the side wheels to prevent mud or dust from being thrown on the rider and to keep clothing from catching.

N. S. C. PERKINS.
Velocipede.

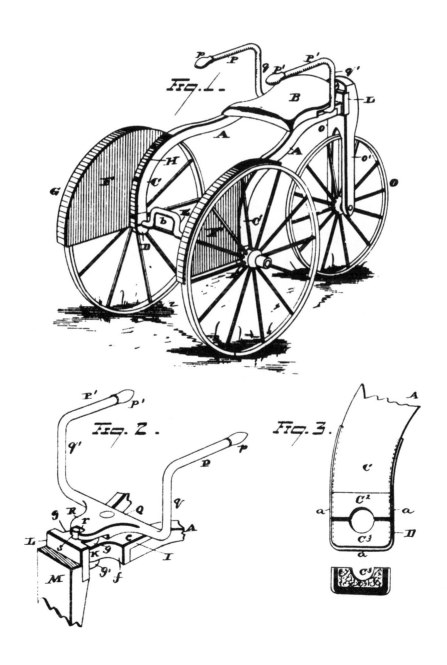

Velocipede. Patent No. 92,528—1869

This vehicle consists of a large wheel with the operator inside. He sits on a saddle with his feet free and his head protected by an awning. The vehicle is propelled by turning handcranks linked to traction equipment fitting inside the rim.

R. C. HEMMINGS.

Velocipede.

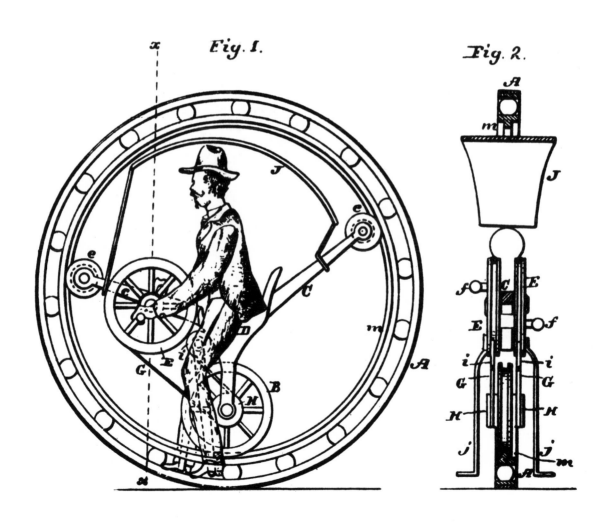

Fig. 1.   Fig. 2.

Velocipede. Patent No. 190,644—1870

This vehicle has a carriage seat for the driver and a hollow front wheel turned by a pair of dogs running inside it.

Traction Engine. Patent No. 221,354—1879

An early type of tractor. Instead of turning wheels the steam engine helped four metal feet take forward steps.

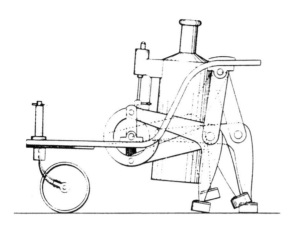

Head Rest. Patent No. 245,639—1881

S. LAY.
HEAD REST.

A drowsy passenger could lay his forehead on a cushioned support *B*, and put his arms on another *C* farther down the pole *A*. The pole rests on the car floor.

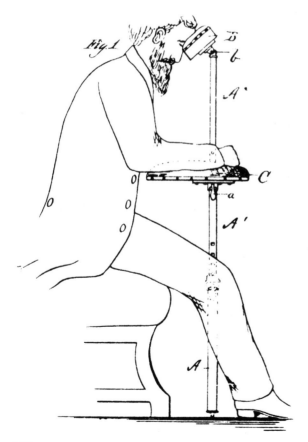

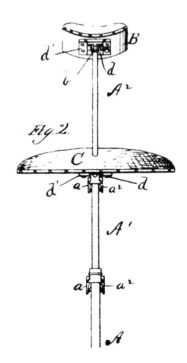

Smoke Conveyors for Locomotives. Patent No. 156,187—1874

Belching smoke and sparks from the engine chimney are funneled past passenger cars and to the rear to protect the passengers.

J. S. THOMAS.

Smoke-Conveyors for Locomotives.

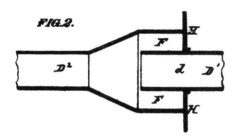

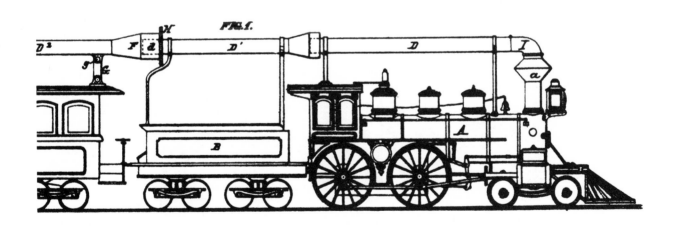

Motor. Patent No. 257,103—1882

An attempt at perpetual motion. Bulb *c* is immersed in water. When filled with air, it rises. This causes pinion *E* to turn gear *F*, making *N* drive end *K* down, moving ball *L* to the right side of the pivot. This enlarges bellows *M*, withdrawing the air from bulb *C*, causing it to sink. This reverses the process until the bellows *M* fills the bulb with air, starting the process all over again.

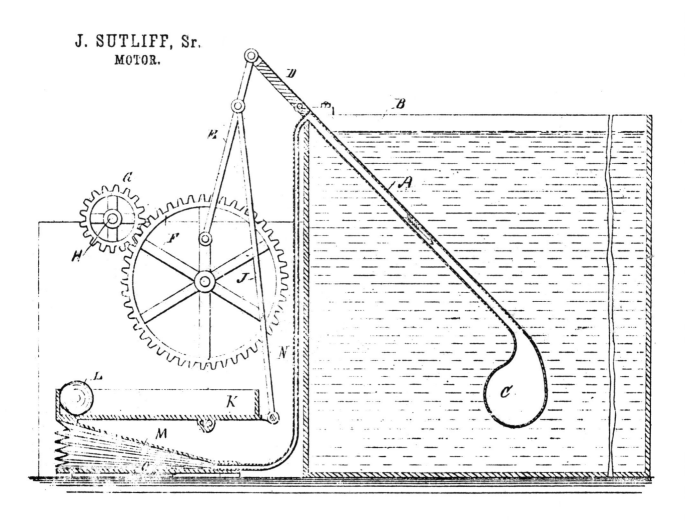

J. SUTLIFF, Sr.
MOTOR.

Railway Gate and Means for Preventing Injury to Stock on Railways. Patent No. 314,990—1885

A lazy-tongs in front of the train has a whistle and water nozzle at the front to frighten animals off the track.

W. BELL.

RAILWAY GATE AND MEANS FOR PREVENTING INJURY TO STOCK ON RAILWAYS.

No. 314,990.

Patented Apr. 7, 1885.

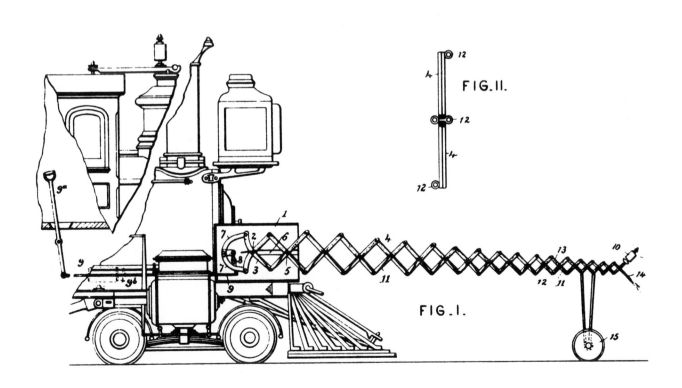

FIG. II.

FIG. I.

Aerial Railway and Car.
Patent No. 328,899—1885

An aerial railway suspended from a cable supported by balloons. The balloons would be raised or lowered to provide the proper downhill angle. The car then moved by gravity.

A. J. MORRISON.
AERIAL RAILWAY AND CAR.

No. 328,899.

Patented Oct. 20, 1885.

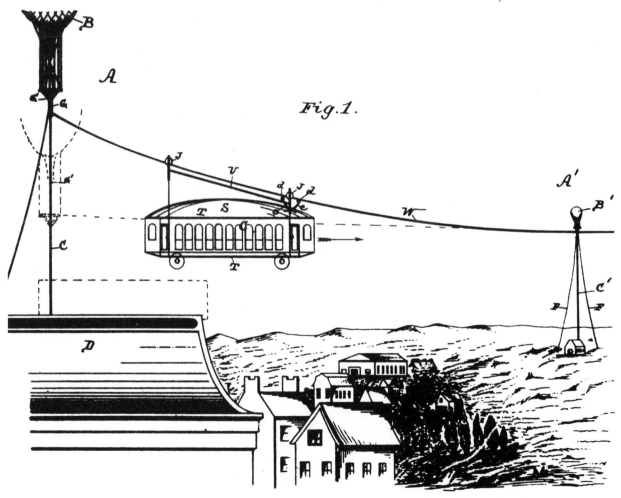

Fig.1.

Means and Apparatus for Propelling and Guiding Balloons. Patent No. 363,037—1887

Harnesses for eagles, vultures and condors are attached to the balloon. A man rotates a cage, causing the birds to fly in a given direction. Rollers direct the flight of the birds up or down.

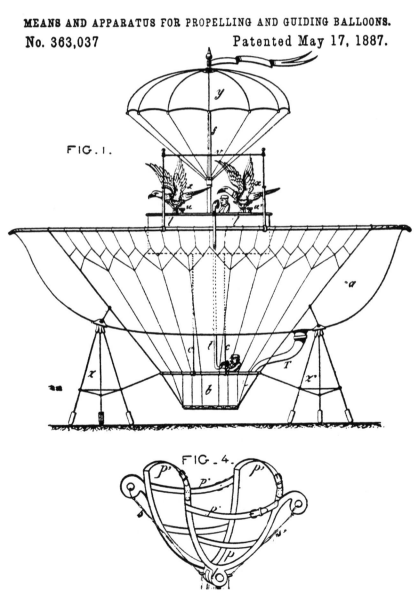

MEANS AND APPARATUS FOR PROPELLING AND GUIDING BALLOONS.

No. 363,037   Patented May 17, 1887.

FIG. 1.

FIG. 4.

Motor for Street Cars. Patent No. 368,825—1887

Putting the horse inside the car was thought to increase speed and decrease cost of operation. The horse faces forward and walks on an endless-belt treadmill. With the horse at an ordinary gait the car can travel from 12 to 18 miles per hour.

Fig. 2.

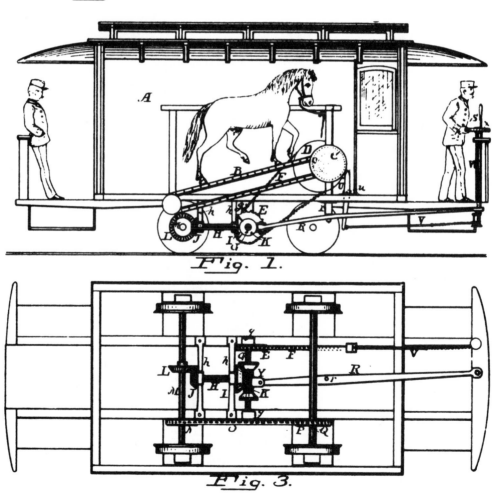

Fig. 1.

Fig. 3.

Aerial Machine. Patent No. 502,168—1893

Although crude, this aerial machine did have elevators and was steerable.

S. B. BATTEY.
AERIAL MACHINE.

Collision Preventer. Patent No. 386,403—1888

A dummy rings a gong on a flatcar telescopically placed ahead of the train engine. This frightens cattle off the track. A pole, supported by wheels, extends in front of the flatcar if the cattle aren't frightened. If the pole strikes an oncoming train, cow, person or other object, an electrical circuit applies the train's air brakes, reverses the engine and pulls in the pilot car. If there is an open bridge ahead, the pilot car falls through first and the train is halted.

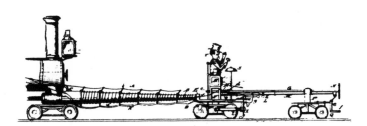

Submarine Vessel. Patent No. 581,213—1897

This is believed to be the first submarine. It has a means for sinking the vessel to the bottom of the water when it is in a state of rest and for raising the vessel to the surface. It travels along the bottom and provides for entry and exit from the vessel when submerged.

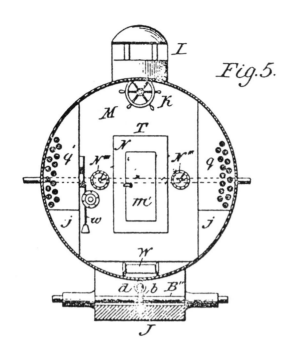

Fig. 5.

S. LAKE.
SUBMARINE VESSEL.

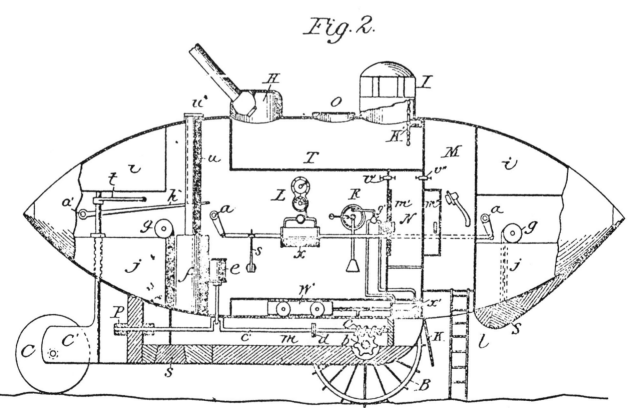

Fig. 2.

Motor Vehicle Attachment.
Patent No. 777,369—1904

A mechanical horse which mounts in front of a car so a real horse will not be disturbed by a "horseless" carriage and will not bolt.

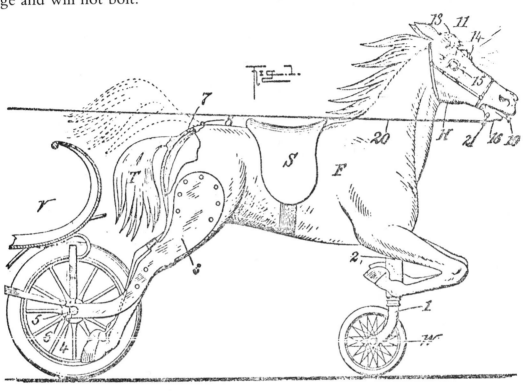

H. HAYES.
MOTOR VEHICLE ATTACHMENT.
APPLICATION FILED MAY 16, 1904.

Message Carrier. Patent
No. 1,469,110—1923

A bottle to carry messages across oceans. The cork is topped by a bell hanging from a support shaped like a question mark. The bell and its support should attract attention.

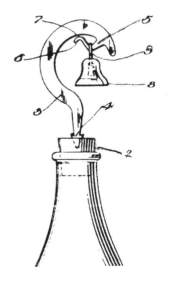

Automobile Attachment.
Patent No. 1,744,727—1930

This tube runs from the driver to a megaphone at the front of the hood. This allows the driver to address persons ahead, facilitating movement of traffic.

Automatic Device for Horseless Vehicles for the Protection of Pedestrians and the Vehicle Itself. Patent No. 1,865,014—1932

This apparatus fits on the front of a car. It stretches a blanket to prevent a person struck by the car from getting hurt by the wheels and to soften his fall.

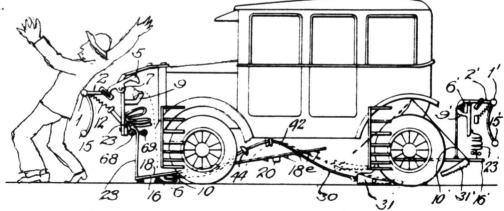

·7·
MEDICAL SCIENCE

No doubt, medical science has, in recent years, benefited from modern technology. Microsurgery, laser surgery, even artificial hearts all help now in saving lives. But using raw meat and electrical voltage to remove poisons? Or wiring a dentist's chair with electricity to anesthetize the patient?

Obviously, the needs that high-tech medical instruments solved in the latter half of the 20th century were still being puzzled over in prior times. Dentistry, because of the pain it usually involved, was a particular enigma, and such inventions as an artificial head for students to practise on (since anesthesia was not available, few were willing to have their mouths examined by students), a tooth extractor that would at least make the operation quick, if not painless, and, as previously mentioned, an electrically wired dentist's chair to "knock out" the patient so the necessary work could be done.

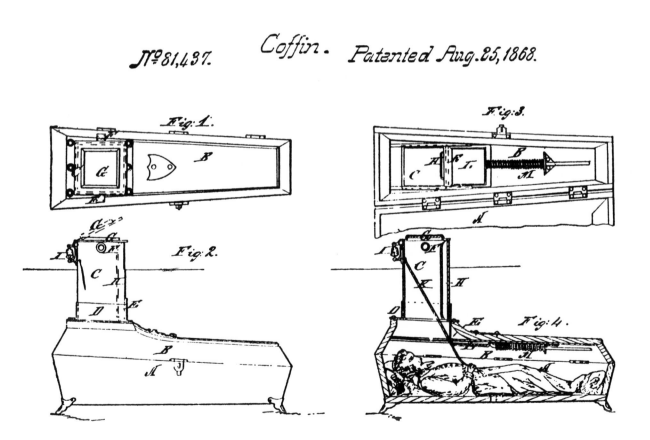

Nº 81,437.   Coffin.   Patented Aug. 25, 1868.

Before the days of modern science, the grisly prospect of being prematurely interred was not very far-fetched. A comatose state was often mistaken for death, and the victim's true condition was only discovered should he or she come around before being lowered into the grave. Thus, the market was seemingly ripe for any device that could prevent a person from dying *in* his own grave, as these four patents show. Patent No. 3,335 assured an escape for the interred by fitting a spring lock to the coffin door that could be worked from within. The door was sufficiently heavy to overturn a shallowly dug grave. This was obviously designed to be used only in the more doubtful cases.

Life-Preserving Coffin. Patent No. 3,335—1843

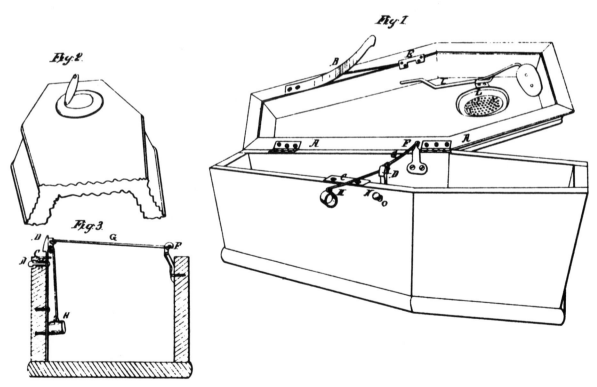

C.H. Eisenbrandt,

Coffin.

N°3,335.

Patented Nov.15,1843.

Coffin. Patent No. 81,497—1868

F. VESTER.
BURIAL CASE.

No. 81,437.   Patented Aug. 25, 1868.

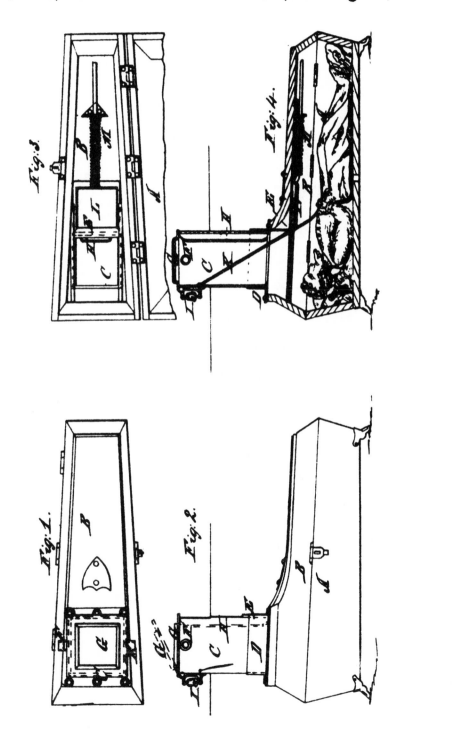

Grave Alarm. Patent
No. 500,072—1893

A. LINDQUIST.
GRAVE ALARM.

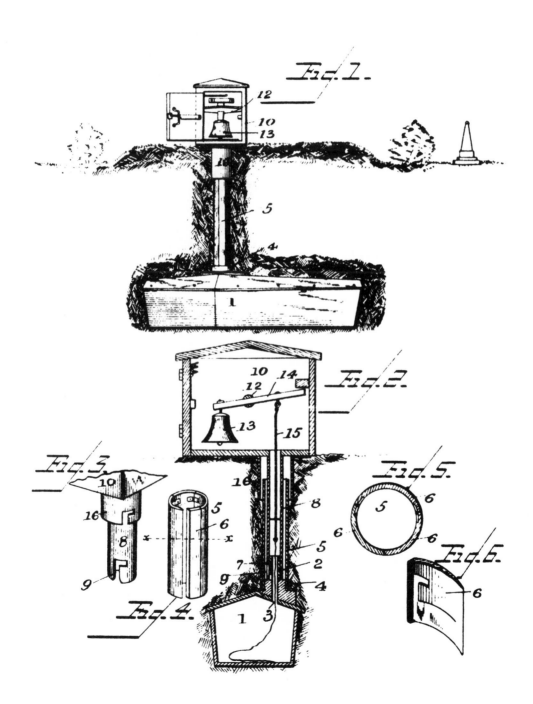

Patent No. 81,497 was only slightly more practical. A square tube, placed over the face of the body and extending to the surface of the grave, contained a ladder and a cord running from the corpse's finger to a bell, which could then be rung for help. If the person really was dead, the tube could be withdrawn, a sliding door closed and the tube reused.

Patent No. 500,072 was another bell-alarm, which would signal that the interred was still alive. On a smaller scale than 81,497, it was permanent, and did not contain a ladder. Variations of this were, in fact, relatively common in some areas of the United States.

Patent No. 901,407 took a different approach, inserting a tube at the foot of the coffin containing two mirrors. Turning a rod adjusted the lower mirror and allowed light to enter. Fresh air was also allowed in through the tube.

Grave Attachment. Patent No. 901,407—1908

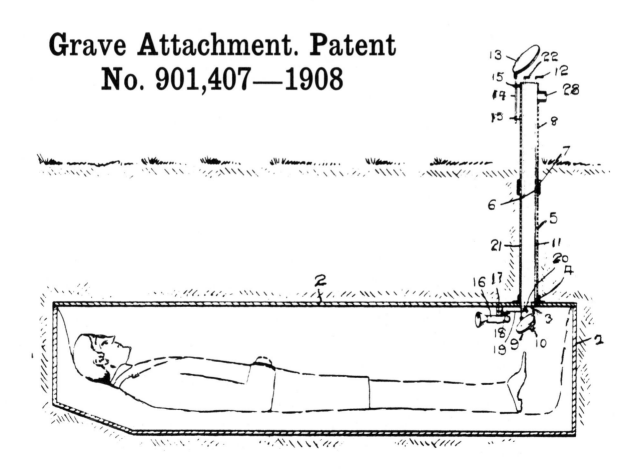

Tooth Extractor—1797

T. BRUFF, Sr.
TOOTH EXTRACTOR.

Patented June 28, 1797.

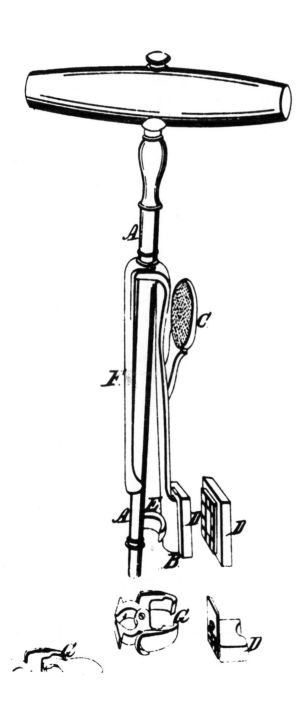

Pads *D* on nearby teeth support the extractor. Rotation of the handle moves claw *E* to pry out the tooth to be extracted.

Electrical Appliance for Dental Chairs. Patent No. 353,403—1886

A dental chair that administered electricity as an anesthetic. A magneto-electric machine behind the chair is operated by the dentist with a pedal sending current to an electrode on each arm of the chair. Current passes through the patient's body when he grasps the arms and the desired shock is produced.

L. L. DECKARD.

ELECTRICAL APPLIANCE FOR DENTAL CHAIRS.

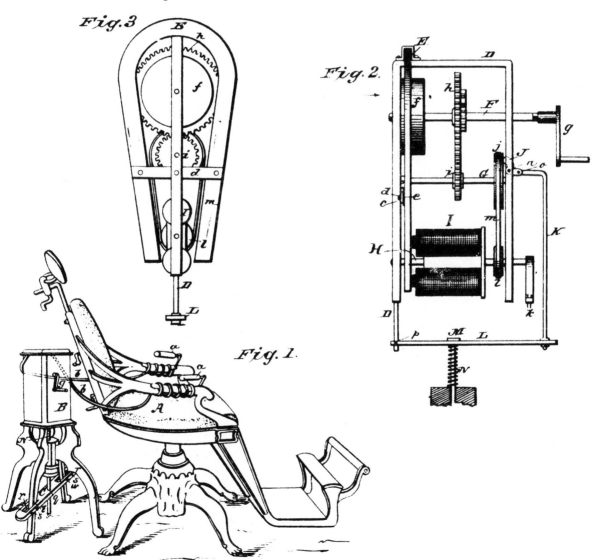

Fig. 3

Fig. 2.

Fig. 1.

Artificial Head for Dentist's Use. Patent No. 451,061—1891

An artificial head for dental students to practise on. It has a metal jaw that can be held closed or open at any angle. The head can be mounted on the back of a dentist's chair in a natural pose. Old human teeth can be fastened in either jaw and removed.

H. C. MAGNUSSON.
ARTIFICIAL HEAD FOR DENTIST'S USE.

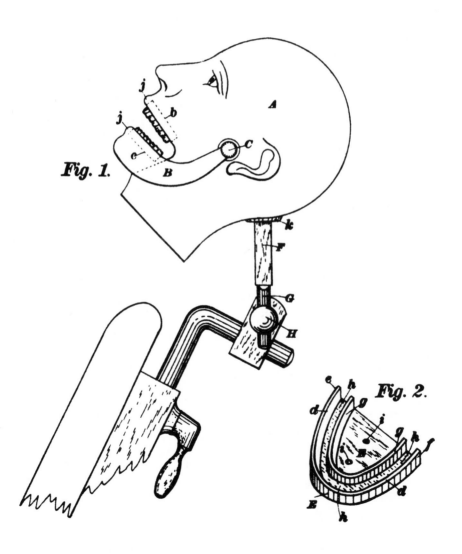

Fig. 1.

Fig. 2.

Electrical Extraction of Poisons. Patent No. 606,887— 1898

The patient sits in chair *1*, electric battery *2* is placed on stand *3* with conducting wires *4* and *5* connected to positive electrode *8* and negative electrode *6* through a copper plate *7* where poisons are to be extracted. The positive electrode *8* is applied to the back of the neck of the patient and the negative electrode copper plate *7* is applied to the patient's bare feet to extract mineral poisons. To extract vegetable poisons, vegetables replace the copper plate, and to extract animal poisons, raw meat is used.

ELECTRIC EXTRACTION OF POISONS.
(Application filed Oct. 5, 1896.)

Method of Preserving the Dead. Patent No. 748,284—1903

A method of preserving the dead by encapsulating the corpse in glass. The corpse is surrounded with a coating of sodium silicate or waterglass, then surrounded with an outer coating of molten glass. The corpse can be viewed in a lifelike state without decomposition or contamination of the viewer.

METHOD OF PRESERVING THE DEAD.
APPLICATION FILED OCT. 13. 1903.

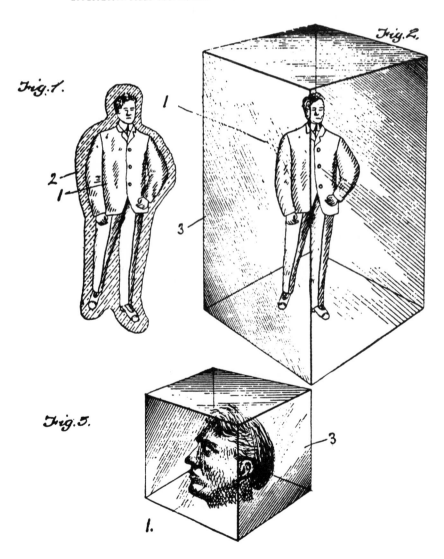

Vertical Casket. Patent No. 3,188,712—1965

This vertical casket is an aluminum cylinder with a ring *21* at the top by which it may be lowered. About five vertical caskets will fit in an area required of one conventional casket. A real space saver.

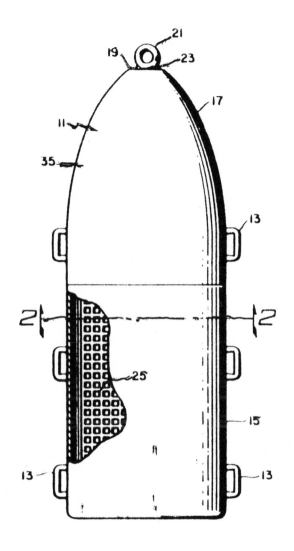

·8·
SPORTS AND GAMES

If patents are based on need, then the need in a sporting event must be tempered by the rules of the game (a patent for a hollow baseball, for example, might fly through the USPO, but not through the Baseball Commissioner's office). Within that parameter, thought, ingenuity and creativity have been put to good use. No doubt many expensive golf clubs were saved by the invention of the pre-broken club, and who knows how many tennis careers could be improved by a rotating racket?

Beyond the world of professional sports, such carnival and circus acts as the human bow-and-arrow and a rigged ball-throwing attraction certainly could keep spirits flying. Fish lures, hunting decoys, and punch counters for boxing matches add variety to this group.

Tennis Racket. Patent No. 335,656—1886

The head of the racket is pivotally adjustable to the handle and neck. This helps return the ball when it bounces close to the ground or over the player's head.

F. W. TAYLOR.

TENNIS RACKET.

No. 335,656.

Patented Feb. 9, 1886.

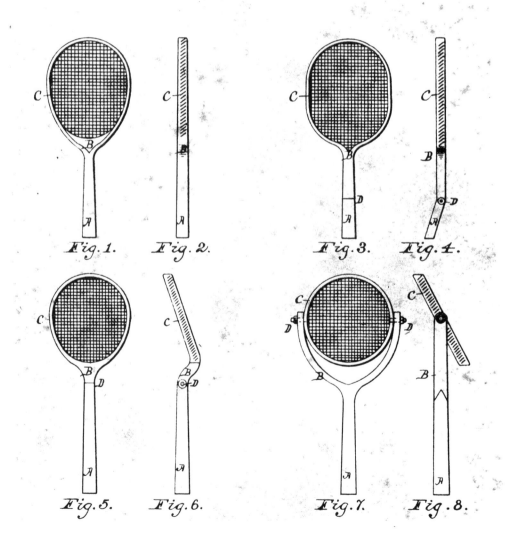

Fig. 1.   Fig. 2.   Fig. 3.   Fig. 4.

Fig. 5.   Fig. 6.   Fig. 7.   Fig. 8.

Fishing Apparatus. Patent No. 515,001—1894

This device induces the fish to take the bait more readily. Bait *S* is put on hook *H* and let down into the water with mirror *A* that serves as a sinker. The fish *B* approaches the bait and sees the reflection in the mirror. The potential companionship or competition will either embolden the fish or make it move closer to take the bait.

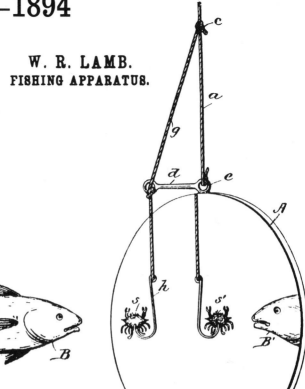

W. R. LAMB.
FISHING APPARATUS.

Registering Armor for Boxers. Patent No. 543,086—1895

The patented armor reduced the roughness and brutality, as well as registered all blows landed. Numbered electrical contacts over the pit of the stomach, heart and short ribs registered hits. Headgear was also padded and wired over the jaws, chin and nose. A bell rang for each blow and a register showed how many blows were landed and where.

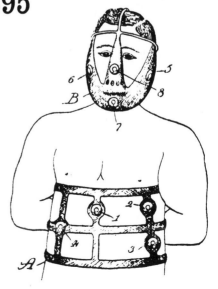

J. DONOVAN.
REGISTERING ARMOR FOR BOXERS.

Hunting Decoy. Patent No. 586,145—1897

A hunter decoy to help hunters bag game. Hunters climb into the cow suit and pull the head and cloth flaps closed. As they roam the fields they fool nearby game.

J. SIEVERS, Jr.
HUNTING DECOY.

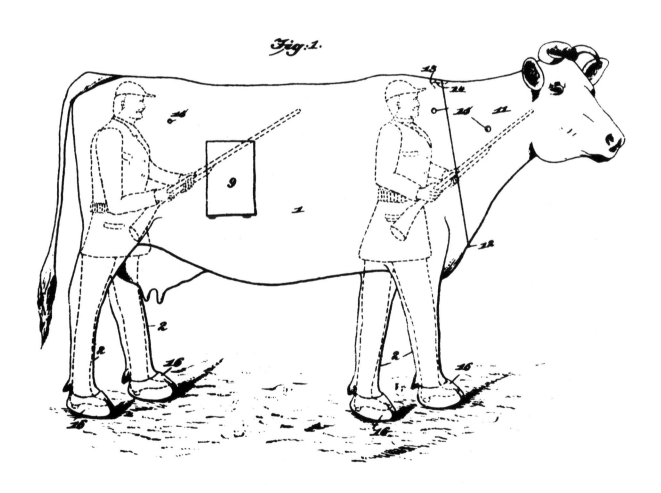

Fig. 1.

Gymnastic Apparatus. Patent No. 562,448—1890

This gymnastic apparatus is a giant bow and arrow on which a human is positioned. When the bow is released, the human is launched, with the arrow dropping off before the human reaches the safety net.

J. ZEDORA.
GYMNASTIC APPARATUS.

Patented June 23, 1896.

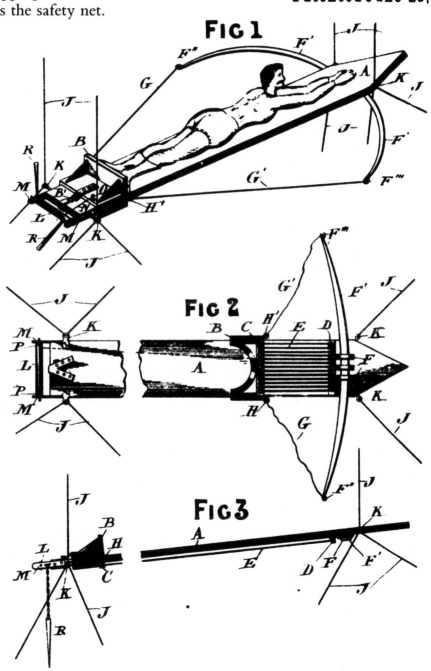

FIG 1

FIG 2

FIG 3

Golf Club. Patent No. 792,631—1905

F. W. TAYLOR.

GOLF CLUB.

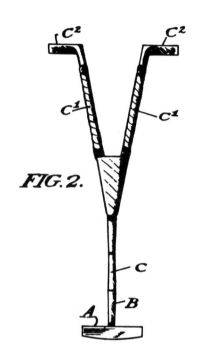

FIG. 2.

This is a two-handled putter. It has a shaft C with two arms C^1 diverging from its center line in such a way that upper arm-rests C^2 bear against the arms of the golfer for stability. The golfer straddles the ball and moves the club like a pendulum between his legs. Mark A^1 on the club head is directly in the golfer's vision, enabling him to align his club during its swinging motion and before it comes in contact with the ball.

Stamp for Simulating Animal Tracks. Patent No. 1,314,276—1919

A device for stamping animal tracks in the snow. After a trapper sets his snare, he obliterates his own footprints and presses into the snow the fake animal impressions. This device may be operated by hand, on the ends of a stick, or attached to wheels alongside a sled.

Amusement Apparatus. Patent No. 1,467,934—1923

A human target lies on his back with his legs in the air and his bottom exposed. A contestant stands on a small platform some distance away and throws baseballs *16*. The man sitting at the target shakes the platform on which the contestant is standing, knocking him off balance and deflecting his aim. The target can also move his body right or left.

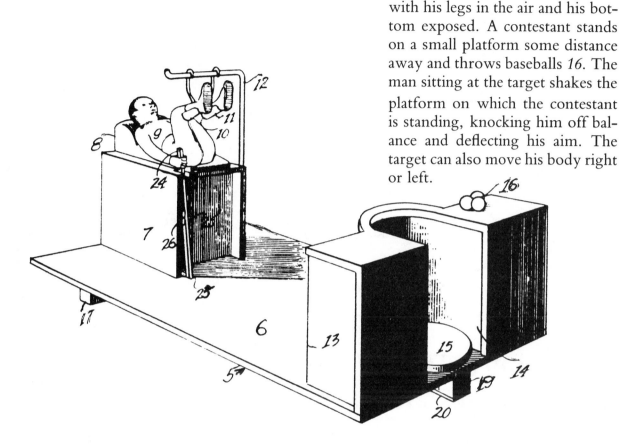

Lost Ball Indicator. Patent No. 1,664,397—1928

When this golf ball comes to rest it releases a visible cloud of ammonium chloride vapor. A coating of phosphorus emits a glow, even in daylight. The scent could also be altered so that dogs could be trained to retrieve the balls.

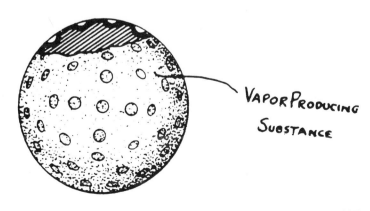

VAPOR PRODUCING SUBSTANCE

Forecasting Calendar. Patent No. 2,595,153—1952

This computer pencil enables a non-professional to weigh and interpret simple horse-racing data. Several sleeves can be rotated to show various statistics. If green shows in all three windows on the pencil barrel, you should bet on that horse, but if red is visible, be careful!

Fish Lure. Design Patent No. 74,759—1927

There are thousands of lures to attract fish into close proximity to the fish hooks attached to them and, hopefully, catch a fish on one of the hooks. Some even work. The lure in this design patent is a mermaid with a woman's body and fishtail legs extending from the waist. A reflective hat is on her head. However, hooks all around make flirtation a hazardous hobby.

Golf Practicing Apparatus.
Patent No. 2,626,151—1953

This machine is designed to prevent a golfer from making a false move. It moves the player's hands, wrists, arms, head, shoulders, hips, knees and feet in one integrated motion, by following the recorded movements of an expert who has preset the machine with his pattern.

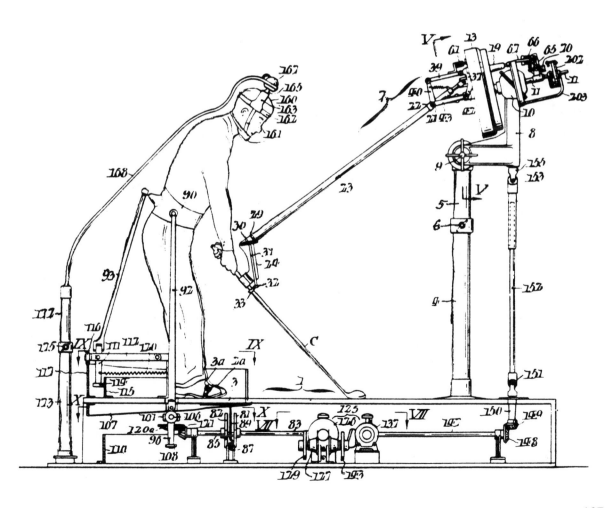

Registering Boxing Glove. Patent No. 2,767,920—1956

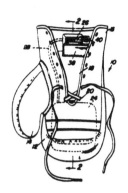

This glove counts the blows, no matter where landed. However, the blows must be of a given force to count.

Game Blind. Patent No. 2,992,503—1961

A game blind that looks like a dead stump. The plastic outer cover resembles tree bark and roots. The top looks like a cut cross-section of the tree. Inside is a float mat and stool, and slots are provided for viewing. Photographers, bird watchers and hunters will find this a comfortable shelter in bad weather.

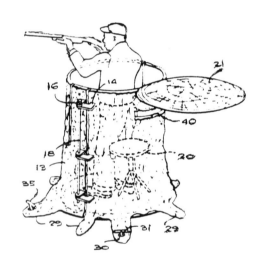

Breakable Simulated Golf Club. Patent No. 3,087,728—1963

After a golfer makes a bad shot, he can take this club from his bag and break the shank. He then replaces the breakable pin and returns the club to the bag. By venting his rage on this club, his regular clubs are not broken.

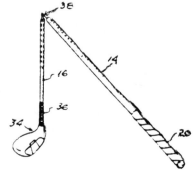

—·9·—
HATS OFF!!

For some reason, hats have held a fascination among inventors and innovators that few other articles of clothing have. Perhaps it is because, more than anything else, a hat tells something about the head that's underneath it.

There are many problems associated with wearing hats, especially for men. How do you keep from sweating on a hot day? Put a ventilator in your chapeau. If your hands are full and you don't wish to be rude when a lady passes, invent a device that will allow the hat to tip itself. And, since men must remove their hats indoors and thus risk having them stolen, a little surprise in the hatband may be enough protection.

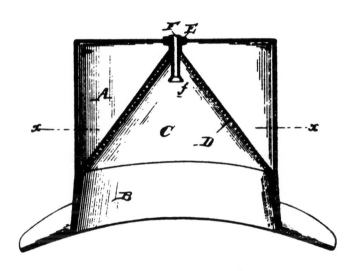

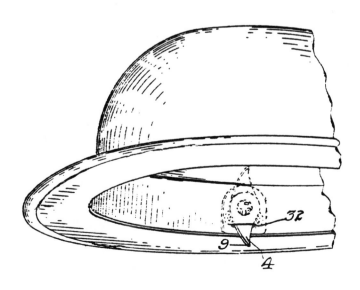

Ventilating Hats. Patent No. 27,985—1860

Hats generate heat, sweating and discomfort. An exhaust or ventilation within the hat makes it more comfortable to wear.

J. Jenkinson.

Ventilating Hats.

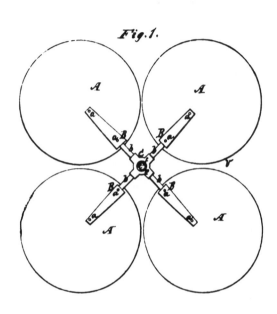

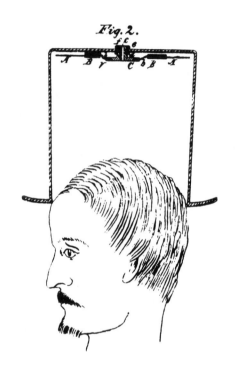

Hats, Cap and Other Head Wear. Patent No. 273,074—1883

A hat, cap or bonnet is dipped in or coated with a self-luminous material that makes it easy to find in dark closets and is beautiful when worn at night. The hat also helps spot a wearer engaged in a hazardous occupation.

R. F. S. HEATH.

HAT, CAP, AND OTHER HEAD WEAR.

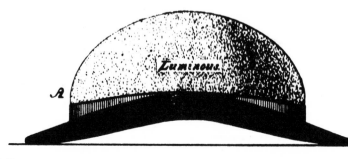

Ear Trumpet. Patent No. 473,608—1892

The wearer who is hard of hearing points the bottom of the hat towards the sound he wants amplified. The small end of the cone terminates in a hole at the top of the hat. The owner places this tube end against his ear.

W. G. A. BONWILL.
EAR TRUMPET.

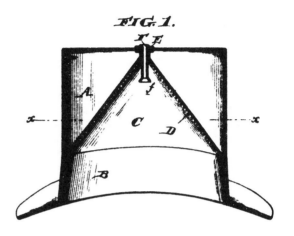

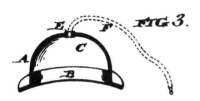

Saluting Device. Patent No. 556,248—1896

This device causes the hat to tip when the wearer bows his head to a person to be saluted. No hands are used.

J. C. BOYLE.
SALUTING DEVICE.

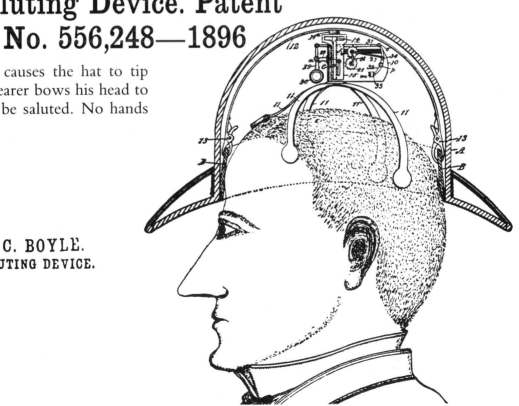

Hat Guard. Patent No. 1,098,691—1914

A hat guard with a sharp-pointed prong 9 in the sweatband. An unauthorized person picking up the hat and putting it on would be painfully jabbed. The owner would know the correct deactivating combination.

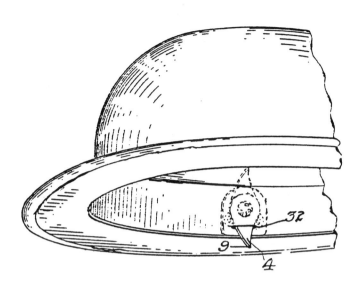

Cooling Unit for a Hat. Patent No. 3,353,191—1967

A solar cell generates current to run the motor. A cover can be swung over the cell to regulate the speed of the fan or shut it off. Air is admitted through holes in the sides of the hat. By cooling the top of the head, the hat cools the entire person.

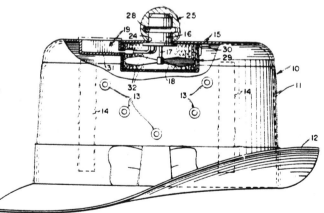

·10·
CRIME AND PUNISHMENT

Crime has been with humankind since Cain and Abel, and fear of crime has provoked the invention of many different kinds of weaponry. Some have been successful, others . . .

The sword was the weapon of choice for many centuries until the firearm was perfected, but for people who couldn't let go of the old ways, why not a combined pistol and sword (when a bayonet just won't do)?

Willie Sutton once said he robbed banks because "that's where the money is." That lure has put banks on alert with guards and hidden cameras. To that one could add exploding money bands or, for the teller who has everything, a pistol that automatically fires when the teller's arms are raised. Hopefully, no one is standing in line behind the robber.

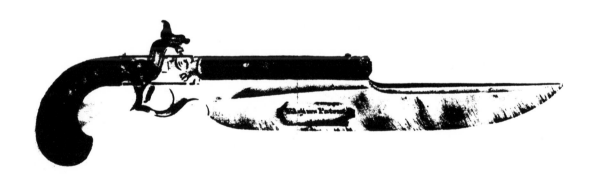

Pistol Sword. Patent No. 254*—1837

A Civil War sidearm, .54 calibre single shot with knife, trigger guard, and hilt all in one piece. The model is in the Smithsonian patent-model-gun collection.

*Note the patent number—254. The patent was issued July 5, 1837. The present numbering system started in 1836 with Patent No. 1 (page 7).

G. ELGIN.
Pistol Sword.

No. 254.   Patented July 5, 1837.

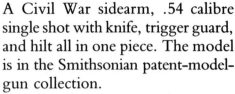

Witnesses:
Wm M Mansell
Tru McCulloh

Inventor
Geo Elgin

Gun Firing Device. Patent No. 1,377,015—1921

A bank cashier (wearing a coat) keeps the pistol *15* under his armpit. A cable *7* extends from the pistol along his arm to a push button *10* in the palm of his hand. In a bank holdup, the cashier lifts his arms and presses the button. This rotates lever *22* counterclockwise to depress trigger *25* and fire the pistol. With luck the cashier is facing the bandit, there is only one of them, and his arm is out of the way of the path of his bullet.

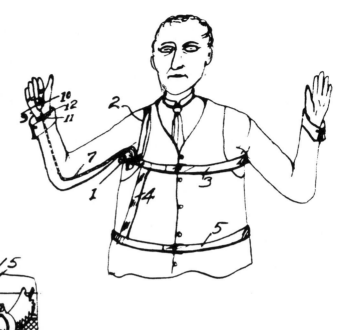

Robbery Protection Device. Patent No. 3,174,245—1965

This robbery protection device is a package of money that explodes when the bandit reaches his getaway car. Stacks of money *1* are wrapped in bands *2*. These stacks have explosive devices hidden inside. A motion detector inside activates the explosive after sufficient time has elapsed for the bandit to reach his car and start his getaway.

—·11·—
AHEAD OF
THEIR TIME

As previously stated, all patents, famous, infamous, or anonymous, started off as strange ideas. However, as humans stand on the brink of the 21st century, some weird ideas are looking less and less so. From major steps like a reuseable rocket launcher (not unlike America's Space Shuttle) to a mouth protector that attaches to a football player's helmet (today in wide use among National Football League players), these ideas have outlasted any premature skepticism and become parts of the modern landscape.

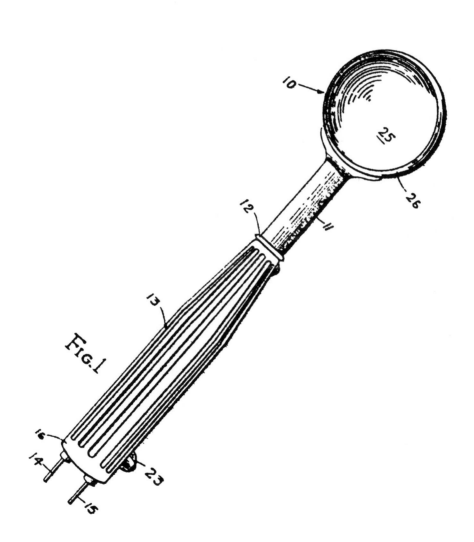

Fig. 1

Coin Controlled Apparatus for Telephones. Patent No. 408,709—1889

The first pay phone was installed in a Hartford, Connecticut bank. The inventor's foreman refused him permission to call his sick wife even when he offered to pay for the call.

W. GRAY.

COIN CONTROLLED APPARATUS FOR TELEPHONES.

No. 408,709.   Patented Aug. 13, 1889.

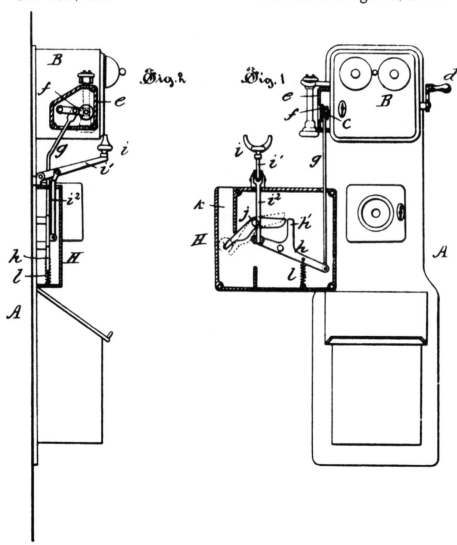

Fragile Article Packaged in Popped Corn. Patent No. 2,049,958—1953

Why not use popcorn to pack articles so they won't break? Practical and environmentally superior to their styrofoam counterparts.

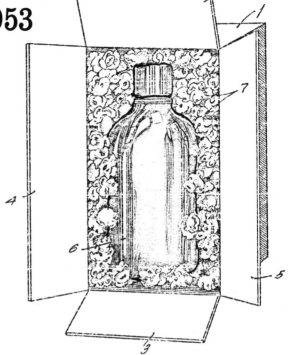

Teeth Protector. Patent No. 3,058,462—1962

A teeth-protecting mouthpiece tied to the face protector of a football helmet not only keeps teeth in place but also prevents lost teeth from dropping onto the ground and allows convenient storage when not in use. This is commonly used by athletes today.

L. L. GREENBLUM

TEETH PROTECTOR

Filed Aug. 23, 1961

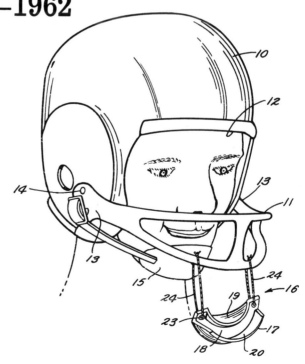

Hoop Toy. Patent No. 3,079,728—1963

It took four years to get this Hula Hoop® patent. However, it was successfully marketed before the patent was issued.

A. K. MELIN

HOOP TOY

Filed May 13, 1959

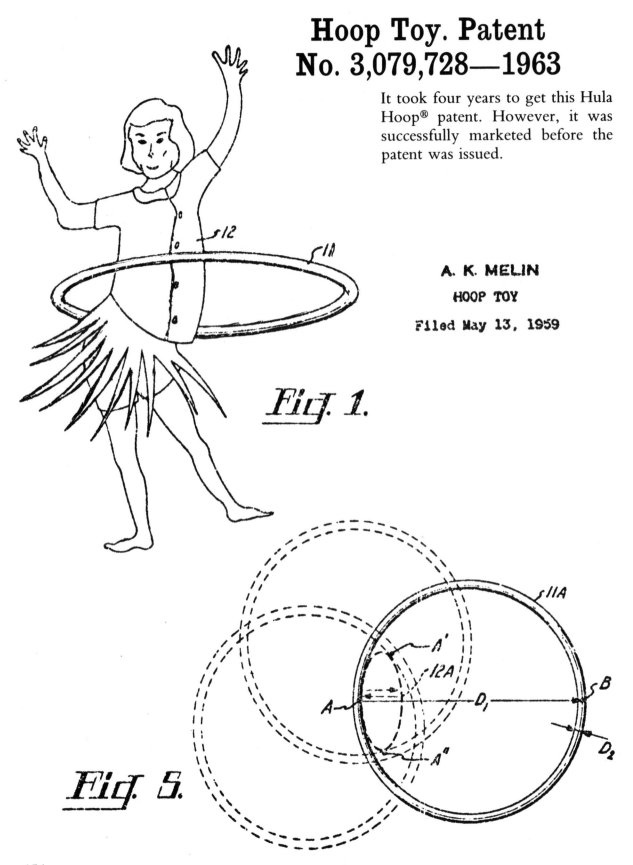

Fig. 1.

Fig. 5.

Shopping Cart Provided with Radio Receiving Apparatus. Patent No. 3,157,871—1964

This improved shopping cart is rugged, light, inexpensive and easy to operate. It also has a radio receiver built in to receive advertising messages from store management. An alarm also sounds when the cart goes beyond a set distance from the store.

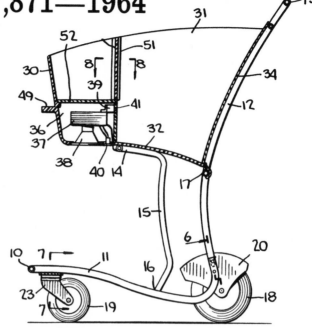

Recoverable Single Stage Spacecraft Booster. Design Patent No. 705,773—1965

The forerunner of the Space Shuttle. This is a design patent relating to the outer appearance of a single stage spacecraft booster operable by the American National Aeronautics and Space Administration.

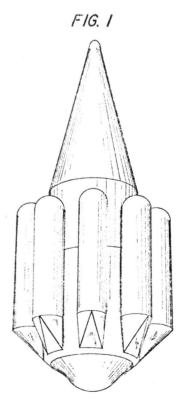

FIG. 1

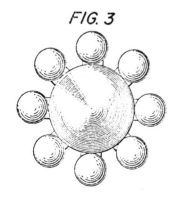

FIG. 3

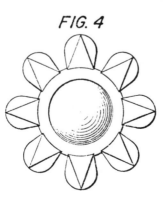

FIG. 4

Automatically Heated Ice-Cream Scoop with Stand. Patent No. 3,513,290—1970

A heated ice-cream scoop can go through and scoop up ice cream easier and faster than a cold one.

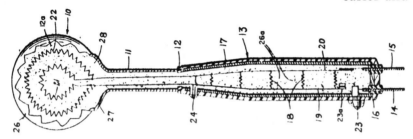

Power Driven File. Patent No. 3,867,747—1975

Before rechargeable electric tools, there was this device. An electric motor and bevel gear convert rotational movement to reciprocal translational motion for the file.

United States Patent [19]

Lee

[54] **POWER DRIVEN FILE**

[76] Inventor: **Robert E. Lee,** 627 N. 20th, Escanaba, Mich. 49829

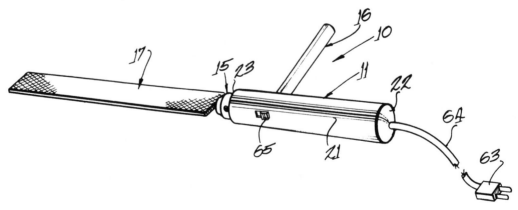

INDEX